SpringerBriefs in Mathematics

SpringerBriefs in Mathematics showcases expositions in all areas of mathematics and applied mathematics. Manuscripts presenting new results or a single new result in a classical field, new field, or an emerging topic, applications, or bridges between new results and already published works, are encouraged. The series is intended for mathematicians and applied mathematicians.

For further volumes:
http://www.springer.com/series/10030

Futaba Fujie • Ping Zhang

Covering Walks in Graphs

Futaba Fujie
Graduate School of Mathematics
Nagoya University
Nagoya, Aichi, Japan

Ping Zhang
Department of Mathematics
Western Michigan University
Kalamazoo, MI, USA

ISSN 2191-8198   ISSN 2191-8201 (electronic)
ISBN 978-1-4939-0304-7   ISBN 978-1-4939-0305-4 (eBook)
DOI 10.1007/978-1-4939-0305-4
Springer New York Heidelberg Dordrecht London

Library of Congress Control Number: 2013957930

Mathematics Subject Classification (2010): 05C12, 05C38, 05C45

Printed on acid-free paper

Springer is part of Springer Science+Business Media (www.springer.com)

Preface

Researchers in graph theory, including graduate students, are the primary audience for this book. Although it is assumed that the reader is acquainted with the basic concepts in graph theory, this book is self-contained in that all concepts and terminology needed for the topic are clearly presented and illustrated. This book can be used for a reading course, a seminar or a short course for graduate students who are interested in Eulerian and Hamiltonian properties of graphs as well as covering walks of graphs in general. In addition, this book contains the background needed to begin a research program on a variety of topics concerning covering walks in graphs and provides easy access to recent results and open problems in this area of research.

Many theorems involving walks in graphs can be traced back to problems that led to some of the best-known and most-studied concepts in graph theory. A walk in a graph G begins at some vertex u of G, proceeds to an edge $e = uu'$ incident with u, then proceeds to u' and next to an edge incident with u' (possibly e again). This continues until the procedure stops at some vertex v, producing a $u - v$ walk W. If W contains every edge of G, then W is an edge-covering walk, while if W contains every vertex of G, then it is a vertex-covering walk.

Graph theory is considered to have begun in 1736 when the great Swiss mathematician Leonhard Euler solved the famous Königsberg Bridge Problem, which asks whether it is possible to walk about the city of Königsberg (located in Prussia at the time) and cross each of its seven bridges in the city exactly once. Eventually it was seen that the Königsberg Bridge Problem could be expressed as a problem in graph theory, an area of mathematics that did not exist in 1736. This led to the concept of Eulerian graphs and later to the more general concept of edge-covering walks in graphs. This is the topic of Chap. 1.

After presenting Euler's characterization of Eulerian graphs as those connected graphs containing only even vertices, graphs containing Eulerian trails are described as those connected graphs containing exactly two odd vertices. Oswald Veblen's characterization of Eulerian graphs as those connected graphs that can be decomposed into cycles is presented. From this theorem, a number of results and conjectures emanated. One of the recent conjectures is the Eulerian Cycle Decomposition Conjecture. Many results obtained on this conjecture are presented.

Several results are presented that deal with connected graphs containing a specified number of odd vertices. While connected graphs with odd vertices do not have circuits containing each edge exactly once, they do contain closed walks containing each edge at least once. Determining the minimum number of edges in such a closed walk is the famous Chinese Postman Problem. This led to the recent study of irregular Eulerian walks in which no two edges are encountered the same number of times in the walk.

Some Eulerian graphs contain vertices u having the property that every trail with initial vertex u can be extended to an Eulerian circuit. Graphs with this property are described. The analogous result for connected graphs with two odd vertices is also presented.

Chapter 2 deals with graphs that possess closed vertex-covering walks. This concept emanates from the so-called Icosian Game of the Irish mathematician William Rowan Hamilton and his Around the World puzzle, which dealt with cycles in a dodecahedron containing every vertex. This led to the concepts of Hamiltonian cycles and Hamiltonian graphs. Although no characterization of Hamiltonian graphs has ever been found, many sufficient conditions for a graph to be Hamiltonian have been discovered, the first of which was a 1952 theorem of the Danish mathematician Gabriel Andrew Dirac, who proved that if the minimum degree of a graph is at least half of the order of the graph, then that graph is guaranteed to be Hamiltonian. This was extended somewhat by the Norwegian mathematician Oystein Ore, who showed that if the sum of the degrees of every two nonadjacent vertices in a graph is at least its order, then that graph is Hamiltonian. This led to the study of the closure of graphs and its connection with Hamiltonian graphs.

The best-known necessary condition for a Hamiltonian graph is one that states that every Hamiltonian graph G has the property that when a set of vertices is removed from G, then the number of components in the resulting graph never exceeds the number of vertices removed. This observation led to the well-studied concept of toughness in graphs and its relationship to Hamiltonian graphs. The famous Traveling Salesman Problem is also described.

Of a number of operations defined on a graph that result in new graphs, two of the most common are the line graph and powers of a graph. There have been numerous theorems dealing with those operations that result in Hamiltonian graphs and graphs with related properties.

Although many connected graphs do not contain Hamiltonian cycles, every connected graph contains a closed vertex-covering walk. The major interest here is the minimum length of such walks in a graph, which is the Hamiltonian number of the graph. Recent research showed that the Hamiltonian number of a graph G can be determined by computing the sum of the distances of consecutive terms in each cyclic ordering of the vertices of G and then finding the minimum of these sums. The maximum of these sums is the upper Hamiltonian number of G and results on this topic are presented as well in Chap. 2. Furthermore, several results on the set of all such numbers obtained in this manner are discussed.

In Chap. 3, the emphasis changes from closed vertex-covering walks in a connected graph G to the recent research topic of open vertex-covering walks in G. Such a walk of minimum length is referred to as a traceable walk and its length is the traceable number of G. Here too it is seen that the traceable number of G

can be obtained by first computing the sum of the distances of consecutive terms in each linear ordering of the vertices of G. The minimum value of such a sum is the traceable number of G; the maximum such sum is the upper traceable number of G. Special attention is given to these two parameters of trees. Also, comparisons between the Hamiltonian and traceable numbers are described as are comparisons between the upper Hamiltonian and upper traceable numbers of given graphs.

One of the differences between closed vertex-covering walks and open vertex-covering walks in a graph is that, for an open vertex-covering walk, its length depends on which vertex the walk begins (or ends). For this reason, for each vertex v in a graph, sequences of the vertices of the graph with v as their initial term and the resulting sums of distances of consecutive terms are considered. The minimum such sum is the traceable number of v. Related concepts such as the maximum vertex-traceable numbers of graphs, traceably singular graphs, and the total traceable numbers of graphs are described.

The numerous new areas of research presented in this book have led to a number of conjectures and open problems, which are described throughout the book.

Nagoya, Aichi, Japan   Futaba Fujie
Kalamazoo, MI, USA   Ping Zhang

Acknowledgements

With pleasure, we thank Gary Chartrand for the advice and information he kindly supplied to us on many topics. In addition, we thank the reviewer for the valuable input and suggestions he/she provided to us. Finally, we are grateful to Razia Amzad, SpringerBriefs editor for her kindness and encouragement in writing this book. It is because of all of you that an improved book resulted.

Contents

List of Figures

Eulerian Walks

1

1.1 The Königsberg Bridge Problem

Not only can the study of covering walks in graphs be traced back to the 1730s but graph theory itself (indeed graphs) can also be traced to the 1730s.

The city of Königsberg was founded in 1255 by the Teutonic knights under the leadership of Bohemian King Ottoker II after the second crusade against the Prussians. Königsberg was the capital of German East Prussia and an important trading city during the Middle Ages due to its location on the banks of the River Pregel. Seven bridges were built across the river, five of which were connected to the island of Kneiphof (see Fig. 1.1). As the map in this figure shows, the river divides Königsberg into four land regions.

The Prussian Royal Castle was located in Königsberg but it was destroyed during World War II, as was much of the city. At the conclusion of the war, it was decided at the Potsdam Conference in 1945 that a region located between Poland and Lithuania, containing the city of Königsberg, would be made part of Russia. In 1946, Königsberg was renamed Kaliningrad after Mikhail Kalinin, the former leader of the Soviet Union during 1919–1946. After the fall of the Soviet Union, Lithuania and other former Soviet republics became independent. Kaliningrad was then no longer part of Russia. Nevertheless, Kaliningrad never had its name changed back to Königsberg.

Early in the eighteenth century, the story goes that many citizens of Königsberg spent their Sunday afternoons strolling about the city. The question arose as to whether it was possible to walk about the city and cross each of the seven bridges exactly once. This problem eventually gained some notoriety and acquired a name known to many:

The Königsberg Bridge Problem. *Is it possible to walk about the city of Königsberg and cross each of its seven bridges exactly once?*

While no one had been able to take such a walk, evidently no one was able to see why such a walk was impossible.

F. Fujie and P. Zhang, *Covering Walks in Graphs*, SpringerBriefs in Mathematics, DOI 10.1007/978-1-4939-0305-4_1, © Futaba Fujie, Ping Zhang 2014

Fig. 1.1 A map of the city
of Königsberg

At this point, Leonhard Euler, one of the brilliant mathematicians of all time, enters the picture. Euler was born in Basel, Switzerland in 1707. It was the well-known Swiss mathematician, Johann Bernoulli, a member of a famous family of mathematicians, who encouraged young Leonhard to study mathematics. Euler became ill while in his 20s, which resulted in his losing vision in one of his eyes. Later he developed a cataract in his other eye and was totally blind during the last few years of his life.

Euler was interested in all areas of mathematics that existed during his time. Among his amazing accomplishments was showing a connection involving perhaps the five best known numbers in mathematics:

$$e^{\pi i} + 1 = 0.$$

He also established an identity involving the number V of vertices, the number E of edges, and the number F of faces in every polyhedron, which became known as the *Euler Polyhedron Formula*:

$$V - E + F = 2.$$

In infinite series, Euler proved in 1735 that

$$\frac{1}{1^2} + \frac{1}{2^2} + \frac{1}{3^2} + \frac{1}{4^2} + \cdots = \frac{\pi^2}{6},$$

an accomplishment that only added to Euler's fame.

Euler was known to be one who corresponded often. Not all individuals with whom he corresponded were mathematicians. One such person was Carl Leonhard Ehler, with whom Euler corresponded during the period 1735–1742. So Ehler's middle name was Euler's first name and his family name differed from Euler's by a single letter. Ehler was the mayor of Danzig in Prussia, a city located some 80 miles west of Königsberg. (Danzig is now the city of Gdansk in Poland.) While there is no clear evidence that Euler first became aware of the Königsberg Bridge Problem through his correspondence with Ehler, these two are known to have discussed the problem in their communications. Ehler was an acquaintance of the mathematician Heinrich Kühn and in a letter dated 9 March 1736, Ehler wrote (in part) the following to Euler:

You would render to me and our friend Kühn a most valuable service, putting us greatly in your debt, most learned Sir, if you would send us the solution, which you know well, to the problem of the seven Königsberg bridges, together with a proof. It would prove to be an outstanding example of the calculus of position worthy of your great genius. I have added a sketch of the said bridges.

The sketch within Ehler's letter showed where the River Pregel flowed through Königsberg and the locations of the seven bridges over the river. It is believed that Ehler asked his question in hopes of promoting the growth of mathematics in Prussia.

Four days after Ehler wrote this letter to Euler, a letter was written by Euler to the Italian mathematician Giovanni Marinoni. In this letter, dated 13 March 1736, Euler described the Königsberg Bridge Problem to Marinoni. Even though Euler didn't consider the problem very difficult, he mentioned to Marinoni that he had solved the problem and explained what there was about this problem that interested him. This letter said in part:

A problem was posed to me about an island in the city of Königsberg, surrounded by a river spanned by seven bridges, and I was asked whether someone could traverse the separate bridges in a connected walk in such a way that each bridge is crossed only once. I was informed that hitherto no-one had demonstrated the possibility of doing this, or shown that it is impossible. This question is so banal, but seemed to me worthy of attention in that not geometry, nor algebra, nor even the art of counting was sufficient to solve it. In view of this, it occurred to me to wonder whether it belonged to the geometry of position, which Leibniz had once so much longed for. And so, after some deliberation, I obtained a simple, yet completely established, rule with whose help one can immediately decide for all examples of this kind, with any number of bridges in any arrangement, whether or not such a round trip is possible ...

In addition to a discussion of the Königsberg Bridge Problem, this letter of Euler contained two additional points of interest. First, Euler referred to a proof technique called the *geometry of position*, originated by the famous German mathematician Gottfried Leibniz, who, with Sir Isaac Newton, was the originator of calculus. In 1670 Leibniz wrote a letter to the prominent Dutch mathematician Christiaan Huygens which said:

I am not content with algebra, in that it yields neither the shortest proofs nor the most beautiful constructions of geometry. Consequently, in view of this, I consider that we need yet another kind of analysis, geometric or linear, which deals directly with position, as algebra deals with magnitude.

Not only was Euler aware of Leibniz's interest in the geometry of position but, as we shall soon see, Euler would do something about it. A second point of interest in Euler's letter to Marinoni was his reference to *round trips*. As we shall also see, Euler distinguished between walks that were round trips and those that were not.

On 3 April 1736, Euler responded to Ehler's letter of 9 March. This letter of Euler to Ehler read in part:

> *Thus you see, most noble Sir, how this type of solution bears little relationship to mathematics, and I do not understand why you expect a mathematician to produce it, rather than anyone else, for the solution is based on reason alone, and its discovery does not depend on any mathematical principle. Because of this, I do not know why even questions which bear so little relationship to mathematics are solved more quickly by mathematicians than by others. In the meantime, most noble Sir, you have assigned this question to the geometry of position, but I am ignorant as to what this new discipline involves, and so to which types of problem Leibniz and Wolff expected to see expressed in this way . . .*

This letter of Euler also referred to Leibniz and his geometry of position as well as the German philosopher Christian Wolff.

Euler showed that it was impossible to stroll about the city of Königsberg and cross each of its seven bridges exactly once, a fact that was probably not surprising to most citizens of Königsberg. Records show that Euler presented this first solution of the Königsberg Bridge Problem in a talk to the members of the Petersburg Academy on 26 August 1735. The following year (in 1736) Euler wrote a paper containing his solution to the Königsberg Bridge Problem. This paper, written in Latin and titled "Solutio Problematis ad Geometriam Situs Pertinentis" (The solution to a problem relating to the geometry of position) was published in the proceedings of the Petersburg Academy (the *Commentarii* [32]).

Euler's paper consisted of 21 numbered paragraphs. He begins his paper by stating that he has been introduced to a problem whose solution uses the geometry of position to which Leibniz had referred. Here we refer to the English translation of Euler's paper presented in the book *Graph Theory 1736–1936* by Biggs, Lloyd, and Wilson [11].

1. *In addition to that branch of geometry which is concerned with magnitudes, and which has always received the greatest attention, there is another branch, previously almost unknown, which Leibniz first mentioned, calling it the geometry of position. This branch is concerned only with the determination of position and its properties; it does not involve measurements, nor calculations made with them. It has not yet been satisfactorily determined what kind of problems are relevant to this geometry of position, or what methods should be used in solving them. Hence, when a problem was recently mentioned, which seemed geometrical but was so constructed that it did not require the measurement of distances, nor did calculation help at all, I had no doubt that it was concerned with the geometry of position – especially as its solution involved only position, and no calculation was of any use. I have therefore decided to give here the method which I have found for solving this kind of problem, as an example of the geometry of position.*

Fig. 1.2 A map of Königsberg with its seven bridges crossing the river Pregel

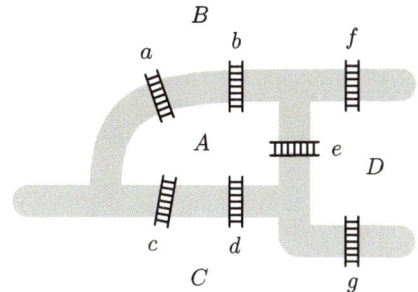

In the second paragraph of this famous paper, Euler states the Königsberg Bridge Problem and says that this has led him to a more general problem.

2. *The problem, which I am told is widely known, is as follows: in Königsberg in Prussia, there is an island A, called the Kneiphof; the river which surrounds it is divided into two branches, as can be seen in Fig. 1.2, and these branches are crossed by seven bridges a, b, c, d, e, f, g. Concerning these bridges, it was asked whether anyone could arrange a route in such a way that he would cross each bridge once and only once. I was told that some people asserted that this was impossible, while others were in doubt: but no one would actually assert that it could be done. From this, I have formulated the general problem: Whatever be the arrangement and division of the river into branches, and however many bridges there be, can one find out whether or not it is possible to cross each bridge exactly once?*

In the third paragraph of his paper, Euler explains how this problem could be approached and why others probably found it so difficult to solve. This paragraph also gives us some insight into Euler's logic.

3. *As far as the problem of the seven bridges of Königsberg is concerned, it can be solved by making an exhaustive list of all possible routes, and then finding whether or not any route satisfies the conditions of the problem. Because of the number of possibilities, this method of solution would be too difficult and laborious, and in the problems with more bridges, it would be impossible. Moreover, if this method is followed to its conclusion, many irrelevant routes will be found, which is the reason for the difficulty of this method. Hence I rejected it, and looked for another method concerned only with the problem of whether or not the specified route could be found; I considered that such a method would be much simpler.*

In paragraphs 4 and 5 of his paper, Euler describes the notation he will use; namely he will be representing the four land regions by A, B, C, D and the seven bridges by a, b, c, d, e, f, g. Also, if a traveler goes from region A to B by either bridge a or b, he will represent this by writing AB and if the traveler should then go to region D, this will be represented by ABD. From this, Euler observes the crossing of all seven bridges requires a sequence of eight letters to represent it.

In the next three paragraphs 6–8, the main observation that Euler makes is that if the number of bridges leading into a region is odd and this number is increased by 1, then half of this result is the number of occurrences of this region in the sequence of eight letters. Thus, as described in paragraph 9, A must occur three times in this sequence and B, C, and D must occur twice each. But this results in nine letters occurring in this sequence of length 8, an impossibility.

In paragraph 10, Euler begins to consider the more general problem – with any number of land regions and any number of bridges. He discusses various possible scenarios throughout paragraphs 10–15. He then writes the following in the next two paragraphs.

16. *In this way it will be easy, even in the most complicated cases, to determine whether or not a journey can be made crossing each bridge once and once only. I shall, however, describe a much simpler method for determining this which is not difficult to derive from the present method, after I have first made a few preliminary observations. First, I observe that the numbers of bridges written next to the letters A, B, C, etc. together add up to twice the total number of bridges. The reason for this is that, in the calculation where every bridge leading to a given area is counted, each bridge is counted twice, once for each of the two areas which it joins.*

17. *It follows that the total of the numbers of bridges leading to each area must be an even number, since half of it is equal to the number of bridges. This is impossible if only one of these numbers is odd, or if three are odd, or five, and so on. Hence if some of these numbers of bridges attached to the letters A, B, C, etc. are odd, then there must be an even number of these. Thus, in the Königsberg problem, there were odd numbers attached to the letters A, B, C, and D ...*

In terms of graphs, the numbers of bridges attached to the letters A, B, C, etc. refer to the degrees of the vertices A, B, C, etc. in the graph representing the situation. What Euler was observing here was a fact that is sometimes called the *First Theorem of Graph Theory* (or the *Handshaking Lemma*), an observation that was then first made by Euler.

Theorem 1.1 (The First Theorem of Graph Theory). *The sum of the degrees of the vertices in a graph equals twice the size of the graph.*

Euler thus observed the following as well. In a graph, a vertex having even (odd) degree is called an *even (odd) vertex*.

Theorem 1.2. *The number of odd vertices in a graph is even.*

In paragraph 20 of his paper, Euler summarizes what he claims to have shown in the foregoing paragraphs.

20. *So whatever arrangement may be proposed, one can easily determine whether or not a journey can be made, crossing each bridge once, by the following rules:*
   - *If there are more than two areas to which an odd number of bridges lead, then such a journey is impossible.*
   - *If, however, the number of bridges is odd for exactly two areas, then the journey is possible if it starts in either of these areas.*
   - *If, finally, there are no areas to which an odd number of bridges lead, then the required journey can be accomplished from any starting point.*
   With these rules, the given problem can also be solved.

Euler concludes his paper with paragraph 21.

21. *When it has been determined that such a journey can be made, one still has to find how
it should be arranged. For this I use the following rule: let those pairs of bridges which
lead from one area to another be mentally removed, thereby considerably reducing
the number of bridges; it is then an easy task to construct the required route across
the remaining bridges, and the bridges which have been removed will not significantly
alter the route found, as will become clear after a little thought. I do not therefore think
it worthwhile to give any further details concerning the finding of the routes.*

While there is no question that in the general situation, a round trip crossing
all bridges exactly once can occur only when the number of bridges leading into
each area is even, a journey between two different areas and crossing each bridge
exactly once can occur only if these two areas are the only areas into which an
odd number of bridges lead. Should there be more than two areas where an odd
number of bridges lead, no journey of any kind is possible that crosses each bridge
exactly once. According to paragraph 21, Euler had evidently convinced himself
that if there is a situation where the number of bridges leading to each area is even,
then the desired round trip is always possible, while if exactly two areas have an odd
number of bridges leading to them, then a desired journey between these is possible.
However, what Euler wrote in these cases was surely not convincing and therefore
not a proof.

While Euler's paper is often credited with being the beginning of graphs and
graph theory, the term *graph* never appeared in his paper. Indeed, the term graph
would evidently not appear (in this context) until 1878 when the British mathematician James Joseph Sylvester first used the word. However, if one were to interpret
each land region as a vertex and each bridge as an edge, graphs did appear abstractly
in Euler's paper and the reasoning that Euler used was graph theoretic in nature.

So, in terms of graphs Euler proved that should a connected graph G contain
more than two vertices of odd degree, then there is no journey about G that contains
every edge of G exactly once. He evidently convinced himself that he had also
verified the converse, namely: If a connected graph G should contain only vertices
of even degree, then there is a round trip in G that contains every edge of G exactly
once – and if G should contain exactly two vertices of odd degree, then there is
a journey about G starting at one of these odd vertices and ending at the other
that contains every edge of G exactly once. But such is not the case. Indeed, a
complete proof of this converse implication did not appear in print for more than
a century later – in 1873 when the young German mathematician Carl Hierholzer
proved this. Sadly, Hierholzer would die before writing a paper that contained his
proof. The paper [42] containing this proof and authored by Hierholzer had the
following footnote written by his colleague C. Wiener:

*Privatdocent Dr. Hierholzer, unfortunately prematurely snatched away by death from the
service of scholarship (died 13 September 1871), reported on the following investigation
to a circle of mathematical friends. It was in order to save it from oblivion (and it had
to be reconstructed without any written record) that I sought to complete the following as
accurately as possible, with the help of my esteemed colleague Lüroth.*

1.2   Eulerian Graphs

As a result of both Euler's work in his famous Königsberg paper as well as
Hierholzer's proof, we are able to state these results as theorems in graph theory.
Prior to doing this, however, it is useful to describe some of the terminology and
notation that we will use. These as well as more basic terminology can be found in
any of the books [19, 20, 24]. All graphs we will be encountering throughout this
book will be nontrivial and connected.

For vertices u and v in a connected graph G, a $u - v$ *walk* W in G is a sequence

$$W = (u = v_0, v_1, v_2, \ldots, v_\ell = v) \tag{1.1}$$

of vertices in G such that $v_{i-1}v_i$ is an edge of G for each i ($1 \le i \le \ell$). If $e_i = v_{i-1}v_i$, then the walk W in (1.1) can also be expressed as

$$W = (e_1, e_2, \ldots, e_\ell). \tag{1.2}$$

The length of the walk W in (1.1) and (1.2) is denoted by $L(W)$, so $L(W) = \ell$.

If G is a multigraph rather than a graph, then some pairs of vertices are joined by
more than one edge. In this case, it is necessary to denote a walk as a sequence of
edges as in (1.2) rather than a sequence of vertices as in (1.1) to avoid confusion. The
$u - v$ walk is *closed* if $u = v$ and *open* otherwise. If there is no repetition of edges in
a walk, then the walk is a *trail*. A closed nontrivial (more than a single vertex) trail
is a *circuit*. A $u - v$ walk W as in (1.1) is a $u - v$ *path* if the vertices $v_0, v_1, \ldots, v_\ell$ are
distinct. If W is a circuit with $L(W) = \ell \ge 3$ for which the vertices $v_0, v_1, \ldots, v_{\ell-1}$
are distinct, then W is a *cycle*.

A circuit in a connected graph G that contains every edge of G is an *Eulerian
circuit*, while an open trail containing every edge of G is an *Eulerian trail*. A
graph containing an Eulerian circuit is an *Eulerian graph* and a graph containing
an Eulerian trail is a *traversable graph*.

A graph is *even* if every vertex in that graph is even. We then have the following.

Theorem 1.3. *A nontrivial connected graph G is Eulerian if and only if G is even.*

Proof. Suppose first that G is an Eulerian graph. Then G contains an Eulerian
circuit C, beginning and ending, say, at u. Let v be a vertex of G different from u.
Then each time we visit v on C, we also leave v. That is, each time v is encountered
on C, a contribution of 2 is made to the degree of v and so $\deg v$ is even. This is
also the case for u should u be encountered in the interior of C. Since C begins and
ends at u, another contribution of 2 is made to the degree of u and so $\deg u$ is even
as well.

For the converse, let G be a nontrivial connected graph in which every vertex has
even degree. Among all trails in G, select one, say T, of maximum length. Suppose
that T is a $u - v$ trail. If $u \ne v$, then T contains an odd number of edges incident with

Fig. 1.3 The Königsberg
multigraph

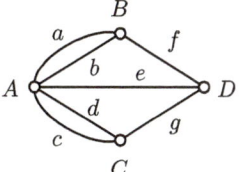

v. Since v has even degree, there is some edge e incident with v and not belonging to T. However, following T by e produces a trail beginning at u that has a greater length than T. Thus, $u = v$ and T is a circuit. Should T contain all edges of G, then G is certainly Eulerian. If this is not the case, then since G is connected, there is some vertex $x \in V(T)$ and an edge $e' = xy \notin E(T)$. However, the edges belonging to T produce an $x - x$ circuit of the same length and so we obtain an $x - y$ trail whose length equals $|E(T)| + 1$ by adding the edge e'. Since this is impossible, T must be an Eulerian circuit and G is Eulerian.   □

An important corollary of Theorem 1.3 is the following.

Theorem 1.4. *A nontrivial connected graph G is traversable if and only if G contains exactly two odd vertices. Any Eulerian trail in G then begins at one of these odd vertices and terminates at the other.*

Proof. If G contains an Eulerian $u-v$ trail, then clearly u and v have odd degree and all other vertices have even degree. For the converse, suppose that G is a connected graph in which only the vertices u and v have odd degree. Let H be the graph obtained by adding a new vertex w to G along with the edges uw and vw. Then H is a connected graph all vertices of which have even degree. Thus H is Eulerian by Theorem 1.3 and so contains an Eulerian circuit C in which uw and vw are consecutive edges on C. Deleting w from C produces a $u - v$ (or $v - u$) Eulerian trail in G and so G is traversable.   □

Both Theorems 1.3 and 1.4 hold for multigraphs as well. Applying these to the multigraph of Fig. 1.3 modeling the four land areas of Königsberg and its seven bridges, we have a solution to the Königsberg Bridge Problem – there is no journey (round trip or otherwise) about Königsberg that crosses each bridge exactly once!

Another characterization of Eulerian graphs is due to the American mathematician Oswald Veblen, one of the early contributors to the field of topology. Veblen helped organize the Institute for Advanced Study in Princeton in 1932 and became the first professor at the Institute. In 1922 Veblen wrote his most influential book, *Analysis Situs* [70], which was what topology was called at that time. This book contained the first systematic coverage of the main principles of topology and contributed to the development of modern topology. While the first book on graph theory was written by Dénes König and published in 1936 [46], Veblen's "topology" book contained some graph theory. Indeed, Chap. 1 of his book was titled *Linear*

Graphs, a term used for graphs at one time. Veblen is also known for his proof of a well-known 1905 theorem of geometry.

The Jordan Curve Theorem. *Every simple closed curve in a plane divides the plane into two disjoint regions, the inside and the outside.*

There are some who believe that Veblen's proof may have been the first rigorous proof of this theorem named for Camille Jordan. Veblen also gave a characterization of Eulerian graphs.

Theorem 1.5 (Veblen's Theorem). [69] *A nontrivial connected graph G is Eulerian if and only if G has a decomposition into cycles.*

Proof. Assume that G has a decomposition $\mathscr{D}$ into cycles. Since G is connected, every vertex belongs to one or more cycles in $\mathscr{D}$ and so its degree is even. Therefore, G is Eulerian.

We now verify the converse, namely that every Eulerian graph has a decomposition into cycles. We proceed by induction on the size $m \geq 3$ of an Eulerian graph. If $m = 3$, then K_3 is the only Eulerian graph of size 3 and is trivially decomposed into a single cycle. For an integer $m \geq 4$, assume that every Eulerian graph of size less than m has a decomposition into cycles. Let G be an Eulerian graph of size m. Then G contains a cycle C. If $G = C$, then G has a trivial cycle decomposition. Otherwise, let H be the graph obtained by deleting edges belonging to C from G. Since the vertices in H are still even, each nontrivial component of H is an Eulerian graph of size less than m and so has a decomposition into cycles. These cycles together with C then produces a cycle decomposition of G.   □

The graph theorist Sabidussi made a conjecture related to Veblen's theorem.

Sabidussi's Conjecture ([34]). *Let G be an Eulerian graph without vertices of degree 2. For each Eulerian circuit C of G, there exists a cycle decomposition $\mathscr{D}$ of G such that every two consecutive edges of C belong to distinct cycles in $\mathscr{D}$.*

When it comes to cycle decompositions, the Eulerian graphs that have received the most attention are the complete graphs of odd order and, to a lesser degree, the complete graphs of even order from which the edges of a 1-factor have been removed (that is, those graphs of even order of the form $K_{2,2,\ldots,2}$). In 1847, the famous mathematician Reverend Thomas Penyngton Kirkman [45] proved that the complete graph of odd order $n \geq 3$ can be decomposed into 3-cycles if and only if its size $\binom{n}{2}$ is divisible by 3. At the other extreme, in 1890 Walecki (see [2]) proved that the complete graph of odd order $n \geq 3$ can always be decomposed into n-cycles. Consequently, when $n \geq 3$ is an odd integer, the complete graph K_n can be decomposed into m-cycles for $m \in \{3, n\}$ if and only if $m \mid \binom{n}{2}$. In fact, Alspach and Jordon [3] proved in 2001 for every pair n, m of odd integers with $3 \leq m \leq n$

that K_n can be decomposed into m-cycles if and only if $m \mid \binom{n}{2}$. In addition, they proved for every even integer $n \geq 4$ and odd integer m with $3 \leq m < n$ that the complete $(n/2)$-partite graph $K_{2,2,\ldots,2}$ can be decomposed into m-cycles if and only if m divides its size $n(n-2)/2$. In 2002 Šajna [60] proved the corresponding results for even integers m. These results verify special cases of a conjecture made by Alspach [1] in 1981.

Alspach's Conjecture. *Suppose that* $n \geq 3$ *is an odd integer and that* $m_1, m_2, \ldots, m_k$ *are integers such that* $3 \leq m_i \leq n$ *for each* i $(1 \leq i \leq k)$ *and* $m_1 + m_2 + \cdots + m_k = \binom{n}{2}$. *Then the complete graph of order* n *can be decomposed into the cycles* $C_{m_1}, C_{m_2}, \ldots, C_{m_k}$. *Furthermore, for every even integer* $n \geq 4$ *and integers* $m_1, m_2, \ldots, m_k$ *such that* $3 \leq m_i \leq n$ *for each* i $(1 \leq i \leq k)$ *with* $m_1 + m_2 + \cdots + m_k = (n^2 - 2n)/2$, *there is a decomposition of the complete* $(n/2)$-*partite graph* $K_{2,2,\ldots,2}$ *into the cycles* $C_{m_1}, C_{m_2}, \ldots, C_{m_k}$.

Alspach's Conjecture was verified in its entirety by Bryant, Horsley, and Pettersson [14] in 2012. A conjecture involving cycle decompositions, introduced in [25], suggests a problem of determining a lower bound on the minimum degree of Eulerian graphs for which the conjecture holds. Mariusz Meszka (personal communication) showed that the following conjecture need not hold for graphs with small minimum degree.

The Eulerian Cycle Decomposition Conjecture (ECDC). *Let G be an Eulerian graph of size* m, *where* a *is the minimum number of odd cycles in a cycle decomposition of G and* c *is the maximum number of odd cycles in a cycle decomposition of G. For every integer* b *satisfying* $a \leq b \leq c$ *and* $b \equiv m \pmod 2$, *there exists a cycle decomposition of G containing exactly* b *odd cycles.*

In the case of the complete graphs of odd order or complete graphs of even order in which a perfect matching has been removed, the maximum number of odd cycles in a cycle decomposition of each such graph is given below. This follows from results of Kirkman [45], Guy [38], and Heinrich, Horák, and Rosa [40].

Theorem 1.6. (a) *For an odd integer* $n \geq 3$, *the maximum number* s *of odd cycles in a cycle decomposition of the complete graph of order* n *is*

$$s = \begin{cases} n(n-1)/6 & \text{if } n \equiv 1, 3 \pmod 6 \\ \lfloor n(n-1)/6 \rfloor - 1 & \text{if } n \equiv 5 \pmod 6. \end{cases}$$

(b) *For an even integer* $n \geq 4$, *the maximum number* s *of odd cycles in a cycle decomposition of the complete* $(n/2)$-*partite graph* $K_{2,2,\ldots,2}$ *is*

$$s = \begin{cases} n(n-2)/6 & \text{if } n \equiv 0, 2 \pmod 6 \\ \lfloor n(n-2)/6 \rfloor - 1 & \text{if } n \equiv 4 \pmod 6. \end{cases}$$

Fig. 1.4 Cycle
decompositions of C_5^2 and C_6^2

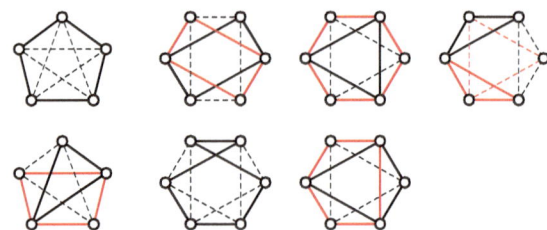

For nontrivial complete graphs of odd order and graphs $K_{2,2,...,2}$ of even order
greater than 2, the ECDC is then a special case of Alspach's Conjecture and
therefore is satisfied for these two classes of graphs.

For a connected graph G and a positive integer k, the k-*th power* G^k of G is that
graph with $V(G^k) = V(G)$ and $E(G^k) = \{uv : 1 \le d_G(u,v) \le k\}$. The graphs G^2
and G^3 are called the *square* and *cube*, respectively, of G.

In a cycle decomposition of an Eulerian graph G, the number of odd cycles in
the decomposition and the size of G are of the same parity. One class of Eulerian
graphs are the squares C_n^2 of cycles C_n where $n \ge 5$. Let C_n^2 be constructed from
$C_n = (v_1, v_2, \ldots, v_n, v_1)$. If n is odd, then the edges in C_n^2 not belonging to C_n form
another n-cycle, namely $(v_1, v_3, \ldots, v_n, v_2, v_4, \ldots, v_{n-1}, v_1)$. Similarly, when n is
even, C_n^2 consists of the two edge-disjoint n-cycles $(v_1, v_2, \ldots, v_{n-2}, v_n, v_{n-1}, v_1)$
and $(v_1, v_3, \ldots, v_{n-3}, v_{n-1}, v_{n-2}, v_{n-4}, \ldots, v_4, v_2, v_n, v_1)$. Thus, C_n^2 has a cycle
decomposition into two n-cycles for each $n \ge 5$.

It is convenient to introduce some notation at this point. If a graph G is
decomposed into a_1 copies of C_{n_1}, a_2 copies of C_{n_2}, and a_3 copies of C_{n_3}, for
example, then we write this decomposition as $G = a_1 C_{n_1} \cup a_2 C_{n_2} \cup a_3 C_{n_3}$. For
example, the above observation shows that C_n^2 always has a decomposition $2C_n$
for $n \ge 5$. Of course, this is not the only cycle decomposition of C_n^2. The graph
C_5^2 has two cycle decompositions, namely $2C_5$ and $2C_3 \cup C_4$. Thus, every cycle
decomposition of C_5^2 has exactly two odd cycles. For C_6^2, there are five cycle
decompositions; namely $3C_4$ and $2C_6$ (no odd cycle), $2C_3 \cup C_6$ and $C_3 \cup C_4 \cup C_5$ (two
odd cycles), and $4C_3$ (four odd cycles). See Fig. 1.4 for these cycle decompositions
of C_5^2 and C_6^2. Note that this is a consequence of Alspach's Conjecture since
$C_5^2 = K_5$ and $C_6^2 = K_{2,2,2}$. In general, we have the following.

Theorem 1.7. *For an integer $n \ge 6$, the minimum number of odd cycles in a
cycle decomposition of C_n^2 is 0 and the maximum number of odd cycles in a cycle
decomposition of C_n^2 is $2\lceil (n+2)/4 \rceil$.*

Proof. Let $G = C_n^2$, where $C_n = (v_1, v_2, \ldots, v_n, v_1)$. We first show that G has a
cycle decomposition in which each cycle is even. Although this holds when n is
even by our earlier observation that G can be decomposed into $2C_n$, we will show
here that G has a cycle decomposition into three even cycles $2C_{2\lceil n/4 \rceil} \cup C_{2(n-2\lceil n/4 \rceil)}$
regardless of the parity of n. Consider the subgraphs G_1 and G_2 of G given by

Fig. 1.5 Decomposing C_n^2 into three even cycles

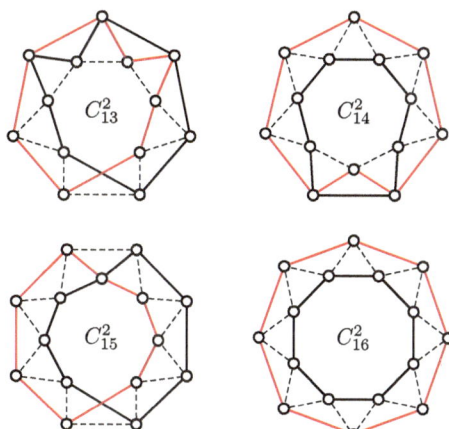

$$G_1 = \begin{cases} (v_1, v_3, v_5, \ldots, v_{n-1}, v_1) & \text{if } n \equiv 0 \pmod 4 \\ (v_1, v_3, v_5, \ldots, v_{n-2}, v_{n-1}, v_n, v_1) & \text{if } n \equiv 1 \pmod 4 \\ (v_1, v_3, v_5, \ldots, v_{n-3}, v_{n-2}, v_n, v_1) & \text{if } n \equiv 2 \pmod 4 \\ (v_1, v_3, v_5, \ldots, v_n, v_1) & \text{if } n \equiv 3 \pmod 4 \end{cases}$$

$$G_2 = \begin{cases} (v_2, v_4, v_6, \ldots, v_n, v_2) & \text{if } n \equiv 0 \pmod 4 \\ (v_2, v_4, v_6, \ldots, v_{n-5}, v_{n-4}, v_{n-3}, v_{n-2}, v_n, v_2) & \text{if } n \equiv 1 \pmod 4 \\ (v_2, v_4, v_6, \ldots, v_{n-2}, v_{n-1}, v_n, v_2) & \text{if } n \equiv 2 \pmod 4 \\ (v_2, v_4, v_6, \ldots, v_{n-1}, v_n, v_2) & \text{if } n \equiv 3 \pmod 4. \end{cases}$$

Observe that G_1 and G_2 are edge-disjoint copies of $C_{2\lceil n/4\rceil}$. Furthermore, there is a cycle decomposition $G = G_1 \cup G_2 \cup G_3$. The third cycle G_3 is also even since its size equals $|E(G)| - (|E(G_1)| + |E(G_2)|) = 2(n - |E(G_1)|)$. (Figure 1.5 shows a cycle decomposition of C_n^2 into $2C_8$ and another even cycle for $13 \le n \le 16$.)

Next, for a given cycle decomposition $\mathscr{D}$ of $G = C_n^2$, let $s_i(\mathscr{D})$ be the number of odd cycles C in $\mathscr{D}$ such that $|E(C) \cap E(C_n)| = i$. Thus, the number of odd cycles in $\mathscr{D}$ equals $\sum_{i \ge 0} s_i(\mathscr{D})$, which we denote by $s(\mathscr{D})$. Note that $s_0(\mathscr{D}) = 0$ if $n \equiv 0 \pmod 4$, $s_0(\mathscr{D}) \le 2$ if $n \equiv 2 \pmod 4$, and $s_0(\mathscr{D}) \le 1$ if n is odd. Also, $\sum_{i \ge 2} s_i(\mathscr{D}) \le \lfloor n/2 \rfloor$. We now consider two cases according to the parity of n.

Case 1. n is even. For $i = 1, 2$, let $G_i = (v_i, v_{i+2}, v_{i+4}, \ldots, v_{i+n-2}, v_i)$. Then we have a cycle decomposition $\mathscr{D}_1$ of G given by $G = G_1 \cup (n/2)C_3$. Since G_1 (as well as G_2) is an $(n/2)$-cycle, it follows that $s(\mathscr{D}_1)$ equals $n/2$ and $n/2 + 1$ if $n \equiv 0 \pmod 4$ and $n \equiv 2 \pmod 4$, respectively. In other words, $s(\mathscr{D}_1) = 2\lfloor (n+2)/4 \rfloor$.

Now we show that $s_1(\mathscr{D}) = 0$ for an arbitrary cycle decomposition $\mathscr{D}$ of G. To see this, consider the graph obtained from G by deleting any $n - 1$ of the n edges belonging to C_n. The resulting graph consists of G_1 and G_2 joined by a bridge. Therefore, there is no cycle (odd or even) containing exactly one edge in C_n, which verifies the claim. Thus, $s(\mathscr{D}) \leq n/2$ if $n \equiv 0 \pmod 4$. For the case $n \equiv 2 \pmod 4$, we see that $s_0(\mathscr{D}) = 2$ if and only if both G_1 and G_2 are contained in $\mathscr{D}$. However then, $\mathscr{D}$ is given by $G = 2C_{n/2} \cup C_n$ and so $s(\mathscr{D}) = 2 < n/2 + 1$. Therefore, $s(\mathscr{D}) \leq n/2 + 1$ for every cycle decomposition $\mathscr{D}$. The result now follows. (In addition, $s(\mathscr{D}) = 2\lfloor(n+2)/4\rfloor$ if and only if $\mathscr{D}$ is given by $G = (n/2)C_3 \cup C_n$.)

Case 2. n is odd. Let $G_3 = (v_1, v_3, \ldots, v_{n-2}, v_n, v_2, v_1)$. Then we obtain a cycle decomposition $\mathscr{D}_2$ of G given by $G = G_3 \cup \lfloor n/2 \rfloor C_3$. Since G_3 is an $(\lceil n/2 \rceil + 1)$-cycle, it follows that $s(\mathscr{D}_2)$ equals $\lfloor n/2 \rfloor$ if $n \equiv 1 \pmod 4$ and $\lceil n/2 \rceil$ if $n \equiv 3 \pmod 4$. Therefore, $s(\mathscr{D}_2) = 2\lfloor(n+2)/4\rfloor$.

Observe that C is a cycle in G satisfying $E(C) \cap E(C_n) = \emptyset$ if and only if $C = (v_1, v_3, \ldots, v_{n-2}, v_n, v_2, v_4, \ldots, v_{n-3}, v_{n-1}, v_1)$, a Hamiltonian cycle in G. If $\mathscr{D}$ is a cycle decomposition of G with $s_0(\mathscr{D}) \geq 1$, then $\mathscr{D}$ contains C and G is decomposed as $G = 2C_n$ and so $s(\mathscr{D}) = 2 < 2\lfloor(n+2)/4\rfloor$. Hence, we next consider cycle decompositions $\mathscr{D}$ of G for which $s_0(\mathscr{D}) = 0$. Suppose that C' is an odd cycle in G such that $|E(C') \cap E(C_n)| = 1$.

Subcase 2.1. $n \equiv 3 \pmod 4$. Then we may assume, without loss of generality, that $C' = G_3$. Furthermore, one sees that no cycle decomposition can contain more than one such odd cycle. Thus, every cycle decomposition $\mathscr{D}$ of G satisfies $s_1(\mathscr{D}) \leq 1$ and so $s(\mathscr{D}) \leq \lfloor n/2 \rfloor + 1 = \lceil n/2 \rceil$. (Hence, $s(\mathscr{D}) = \lceil n/2 \rceil$ if and only if $\mathscr{D}$ is given by $G = \lfloor n/2 \rfloor C_3 \cup C_{\lceil n/2 \rceil + 1}$.)

Subcase 2.2. $n \equiv 1 \pmod 4$. Then we may assume that $C' = (v_1, v_2, v_4, v_6, \ldots, v_{n-1}, v_1)$, which is an $\lceil n/2 \rceil$-cycle. If a cycle decomposition $\mathscr{D}$ containing C' has another such odd cycle C'', then we may again assume that $C'' = (v_1, v_3, v_5, \ldots, v_n, v_1)$. However then, $\mathscr{D}$ is given by $G = 2C_{\lceil n/2 \rceil} \cup C_{n-1}$. Therefore, if $s_1(\mathscr{D}) \geq 2$, then $s_1(\mathscr{D}) = s(\mathscr{D}) = 2$. This allows us to assume that $s_1(\mathscr{D}) \leq 1$. If $s_1(\mathscr{D}) = 1$, say $\mathscr{D}$ contains C', then there are $3\lfloor n/2 \rfloor + 1$ edges not belonging to C', four of which (namely the ones incident with v_1 or v_2) cannot belong to a triangle in $\mathscr{D}$. This implies that $\sum_{i \geq 2} s(\mathscr{D}) \leq \lfloor n/2 \rfloor - 1$ and we obtain $s(\mathscr{D}) \leq \lfloor n/2 \rfloor$. (In this case, $s(\mathscr{D}) = \lfloor n/2 \rfloor$ if and only if $\mathscr{D}$ is given by either $G = \lfloor n/2 \rfloor C_3 \cup C_{\lceil n/2 \rceil + 1}$ or $G = (\lfloor n/2 \rfloor - 1) C_3 \cup C_4 \cup C_{\lceil n/2 \rceil}$.) □

In general, if $n \geq 5$ is odd, then $C_n^2 = 2C_n$; if $n \equiv 2 \pmod 4$, then $C_n^2 = 2C_{n/2} \cup C_n$; while if $n \equiv 0 \pmod 4$, then $C_n^2 = C_{n/2} \cup C_{n/2+1} \cup C_{n-1}$. Therefore, C_n^2 has a cycle decomposition with exactly two odd cycles for each integer $n \geq 5$. Similarly, $C_7^2 = 3C_3 \cup C_5$ while for every $n \geq 6$ $(n \neq 7)$ the graph C_n^2 has a cycle decomposition into four triangles and even cycles; so C_n^2 has a cycle decomposition with exactly four odd cycles for $n \geq 6$. In [25], the ECDC was verified for Eulerian complete 3-partite graphs and Eulerian k-th powers of cycles for $k = 2, 3, 4$. For an odd integer $n \geq 3$, we have $C_n^{\lfloor n/2 \rfloor} = K_n$. Therefore, the maximum number of odd cycles in a cycle decomposition of C_n^k is known for $k \in \{1, 2, 3, 4, \lceil n/2 \rceil - 1\}$. For an even integer $n \geq 4$, we have $C_n^{n/2-1} = K_{2,2,\dots,2}$ (and $C_n^{n/2} = K_n$). Therefore, the maximum number of odd cycles in a cycle decomposition of C_n^k is known for $k \in \{1, 2, 3, 4, \lceil n/2 \rceil - 1\}$ in this case as well. Furthermore, the ECDC is true for each of these graphs as well as many other complete multipartite graphs.

We have now seen a characterization of Eulerian graphs as those connected graphs in which every vertex is incident with an even number of edges as well as those connected graphs having a decomposition into cycles. Another characterization of Eulerian graphs that concerns both parity and cycles, jointly due to Toida [64] and McKee [51], is stated next. The necessity is due to Toida and the sufficiency to McKee.

Theorem 1.8. *A nontrivial connected graph G is Eulerian if and only if every edge of G lies on an odd number of cycles in G.*

Proof. Suppose first that $e = uv$ is an edge in an Eulerian graph G. Since e is not a bridge, $G - e$ contains $u - v$ trails. Let us consider the set $\mathscr{S}$ of those $u - v$ trails in which v appears exactly once. When constructing a trail $T \in \mathscr{S}$ with initial vertex u, there is an odd number of choices for the first edge e_1 of T. Once e_1 has been selected, there is again an odd number of choices for the next edge of T. This procedure is continued until the trail arrives at v and T is obtained. As a consequence, $|\mathscr{S}|$ is odd.

While every $u - v$ path in $G - e$ is contained in $\mathscr{S}$, there may be some trails in $\mathscr{S}$ that are not paths. Let $\mathscr{S}'$ be the subset of $\mathscr{S}$ containing those trails that are not paths. If $T \in \mathscr{S}'$, one can write $T = (u = v_0, v_1, v_2, \dots, v_\ell = v)$, where $\ell = L(T)$ and $v_{i_1} = v_{i_2}$ for some integers i_1, i_2 satisfying $2 \leq i_1 + 2 \leq i_2 \leq \ell - 1$. Thus, T contains a circuit $C = (v_{i_1}, v_{i_1+1}, v_{i_1+2}, \dots, v_{i_2} = v_{i_1})$. Then there is a corresponding trail T' obtained from T by replacing C by the circuit $(v_{i_1} = v_{i_2}, v_{i_2-1}, v_{i_2-2}, \dots, v_{i_1})$. Since T' also belongs to $\mathscr{S}'$, it follows that the trails in $\mathscr{S}'$ occur in pairs, that is, $|\mathscr{S}'|$ is even. Therefore, the number of $u - v$ paths in $G - e$, which equals the number of cycles containing e in G, must be odd.

We now verify the converse. Let G be a nontrivial connected graph in which every edge lies on an odd number of cycles in G. For each vertex v, let $\{e_1, e_2, \dots, e_{\deg v}\}$ be the set of edges incident with v. If s_i denotes the number of cycles in G containing the edge e_i for $1 \leq i \leq \deg v$, then $\sum_{i=1}^{\deg v} s_i$ must be even, since it counts each cycle containing v twice. It follows that $\deg v$ is even since each s_i is odd. Therefore, G is Eulerian. $\square$

We have mentioned that every Eulerian graph has a cycle decomposition. Indeed, by Theorem 1.5, Eulerian graphs are characterized as those connected graphs possessing a cycle decomposition. Two cycle decompositions of a *labeled* Eulerian graph G are considered different if these two decompositions do not consist of exactly the same cycles. For example, while we have seen that the complete graph of order 5 has two cycle decompositions $2C_5$ and $2C_3 \cup C_4$, when the vertices are labeled, there are actually 21 distinct cycle decompositions as shown in Fig. 1.6.

Fig. 1.6 The 21 cycle decompositions of the labeled K_5

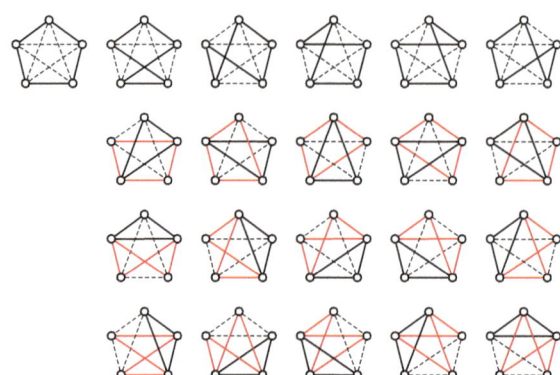

Bondy and Halberstam [13] found a more specialized characterization of Eulerian graphs into cycle decompositions. One of its consequences is that there is no labeled graph having an even number of cycle decompositions unless the graph has no cycle decomposition at all.

Theorem 1.9. *A labeled connected graph G is Eulerian if and only if G has an odd number of cycle decompositions.*

Proof. If G has an odd number of cycle decompositions, then G has at least one cycle decomposition and so G is Eulerian by Theorem 1.5.

We verify the converse by induction on the size $m \geq 3$ of Eulerian graphs. For $m = 3$, the result is immediate as K_3 is the only Eulerian graph of size 3, which clearly has exactly one cycle decomposition. For an integer $m \geq 4$, assume that all Eulerian graphs of size less than m have an odd number of cycle decompositions. Let G be an Eulerian graph of size m. If G is an m-cycle, then G has exactly one cycle decomposition. Thus, assume that G is not a cycle.

Let $e \in E(G)$ and suppose that C is a cycle containing e. We now consider the number s of cycle decompositions of the graph H obtained from G by deleting the edges belonging to C. Certainly every nontrivial component of H is Eulerian. Thus, if H is connected, then s is odd by the induction hypothesis. Otherwise, suppose that $H_1, H_2, \ldots, H_k$ are the nontrivial components of H and they have $s_1, s_2, \ldots, s_k$ respective cycle decompositions. Since each s_i is odd again by the induction hypothesis, so is $s = \prod_{i=1}^{k} s_i$. Each cycle decomposition of H together with C results in a cycle decomposition of G and we have shown that there are s of them, where s is odd. Since the number of cycles containing e is odd by Theorem 1.8, it follows that the total number $\sum s$ of cycle decompositions of G is also odd.   □

Table 1.1 The values of
$f_1(G)$ and $f_2(G)$

G	C_4	C_5	$K_{2,3}$	$K_{1,1,2}$
$f_1(G)$	5	6	19	14
$f_2(G)$	15	31	54	24

Vector Spaces Associated with a Graph

Eulerian graphs have another interesting property in terms of the number of certain subgraphs they contain. For a connected graph G, observe that H is a connected spanning subgraph of G if and only if there exists a spanning tree T of G such that $T \subseteq H$; while H is an acyclic subgraph of G if and only if there exists a spanning tree T of G such that $H \subseteq T$. Suppose, for G, that we define f_i ($i = 1, 2$) by

$$f_1(G) = \text{ the number of connected spanning subgraphs of } G$$

$$f_2(G) = \text{ the number of acyclic subgraphs of } G.$$

For example, let us consider the four graphs C_4, C_5, $K_{2,3}$, and $K_{1,1,2}$. The values of f_i ($i = 1, 2$) for these graphs are summarized in Table 1.1. For these four graphs, we see that $f_1(G)$ is odd if and only if G is bipartite and $f_2(G)$ is odd if and only if G is Eulerian, which is in fact the case for connected graphs in general.

For two arbitrary sets X and Y, the *symmetric difference* $X \triangle Y$ of X and Y is defined by $X \triangle Y = (X \cup Y) \backslash (X \cap Y)$. Observe that (i) $X \triangle Y = Y \triangle X$, (ii) $X \triangle X = \emptyset$, (iii) $X \triangle \emptyset = \emptyset \triangle X = X$. Also,

$$|X \cap (Y \triangle Z)| \equiv |X \cap Y| + |X \cap Z| \pmod 2 \tag{1.3}$$

for any three sets X, Y, Z since $|X \cap (Y \triangle Z)| = |X \cap Y| + |X \cap Z| - 2|X \cap Y \cap Z|$.

The facts (i)–(iii) imply that the power set $\mathscr{P}_A$ of an arbitrary set A together with the binary operation $\triangle$ is an abelian group in which $\emptyset$ is the identity and every element is its own inverse. In fact, this is a vector space over the field $\mathbb{F}_2 = \{0, 1\}$. When A is finite, $\dim \mathscr{P}_A = |A|$.

Suppose that A is a finite set and consider a subspace $\mathscr{W}$ of $\mathscr{P}_A$. Then the number of elements in $\mathscr{W}$ equals $2^{\dim \mathscr{W}}$. If we define

$$\mathscr{W}_0 = \{X \subseteq A : |X \cap Y| \equiv 0 \pmod 2 \text{ for all } Y \in \mathscr{W}\},$$

then $\mathscr{W}_0$ is another subspace of $\mathscr{P}_A$ by (1.3) and contains $2^{\dim \mathscr{W}_0}$ elements. Note that $\mathscr{W} \cap \mathscr{W}_0$ may or may not be trivial and $\dim \mathscr{W} + \dim \mathscr{W}_0 \geq \dim \mathscr{P}_A$ in general.

The following result by Shank [62] is stated and verified in a slightly different way here.

Theorem 1.10. *For a finite set A, let $\mathscr{W}$ be a subspace of $\mathscr{P}_A$. The number of subsets of A that meet every nonempty element in $\mathscr{W}_0$ is odd if and only if $A \in \mathscr{W}_0$ (that is, the cardinality of each set in $\mathscr{W}$ is even).*

Proof. Let

$$U = \{X \in \mathscr{P}_A : X \cap Y \neq \emptyset \text{ for all } Y \in \mathscr{W}_0 \backslash \{\emptyset\}\}$$
$$V = \{(X, Y) \in \mathscr{P}_A \times \mathscr{W}_0 : X \cap Y = \emptyset\}.$$

Let X be a fixed subset of A. Then $\mathscr{W}_{0,X} = \{Y \in \mathscr{W}_0 : Y \cap X = \emptyset\}$ is a subspace of $\mathscr{W}_0$. Thus, $|\mathscr{W}_{0,X}| \equiv 1 \pmod 2$ if and only if $X \in U$ and so

$$|V| = \sum_{X \in \mathscr{P}_A} |\mathscr{W}_{0,X}| \equiv |U| \pmod 2. \qquad (1.4)$$

On the other hand, $|\mathscr{P}_{A\backslash X}| = 2^{|A|-|X|} \geq 1$ and $|\mathscr{P}_{A\backslash X}| \equiv 1 \pmod 2$ if and only if $X = A$, which implies that

$$|V| = \sum_{X \in \mathscr{W}_0} |\mathscr{P}_{A\backslash X}| \equiv \begin{cases} 0 \pmod 2 & \text{if } A \notin \mathscr{W}_0 \\ 1 \pmod 2 & \text{if } A \in \mathscr{W}_0. \end{cases} \qquad (1.5)$$

By (1.4) and (1.5), therefore, $|U|$ is odd if and only if $A \in \mathscr{W}_0$. $\qquad \square$

For a graph G, the power set $\mathscr{E}(G) = \mathscr{P}_{E(G)}$ of its edge set forms a vector space (in which the addition of two vectors is their symmetric difference), called the *edge space* of G.

The subspace $\mathscr{C}$ of $\mathscr{E}(G)$ spanned by the subsets of $E(G)$ that induce cycles in G is called the *cycle space* of G and its dimension equals $|E(G)| - |V(G)| + 1$. For example, by selecting a fixed spanning tree T of G, each of the $|E(G)| - |V(G)| + 1$ edges not contained in T produces a cycle in G. The edge sets of these cycles form a basis of $\mathscr{C}$ (called a *cycle basis*). Observe that $X \in \mathscr{C}$ if and only if the subgraph induced by X is even.

For a connected graph G, a set $X \subseteq E(G)$ is an *edge-cut* of G if $G - X$ is disconnected. If X is an edge-cut of G, then so is Y whenever Y satisfies $X \subseteq Y \subseteq E(G)$. If X is an edge-cut of G while no proper subset of X is, then X is called a *minimal edge-cut* of G. When X is a minimal edge-cut of G, the graph $G - X$ consists of exactly two components. Every edge-cut of G contains a minimal edge-cut of G as a subset.

Lemma 1.1. *For a nontrivial connected graph G with the cycle space $\mathscr{C}$, every nonempty element of $\mathscr{C}_0$ is an edge-cut of G inducing a bipartite subgraph. In particular, every minimal edge-cut of G belongs to $\mathscr{C}_0$.*

Proof. Suppose that X is a nonempty subset of $E(G)$. First, if $G[X]$ contains a cycle C, then $E(C) \in \mathscr{C}$ and $|X \cap E(C)| = |E(C)|$. Thus, if $X \in \mathscr{C}_0$, then $G[X]$ cannot contain odd cycles. Next, if $G - X$ is connected, then let uv be an edge in

X. Then $G - X$ contains a $u - v$ path P and so $P + uv$ is a cycle in G containing exactly one edge of X. Since $E(P + uv) \in \mathcal{C}$, it follows that $X \notin \mathcal{C}_0$.

Let X be a minimal edge-cut of G. Then $|X \cap E(C)|$ is even (and $|X \cap E(C)| = 0$ if and only if C is entirely contained in one of the two components in $G - X$) for each cycle C in G. For each nonempty $Y \in \mathcal{C}$, every component of $G[Y]$ has a cycle decomposition by Theorem 1.5, which then implies that $|X \cap Y|$ is even. $\square$

A *cut* is a partition of $V(G)$ into two nonempty subsets, say V_1 and V_2. The set of all edges joining a vertex in V_1 and a vertex in V_2 is called a *cocycle*. Hence, X is a cocycle if and only if X induces a bipartite subgraph. In particular, every element in $\mathcal{C}_0$ is a cocycle by Lemma 1.1. Suppose that $V(G) = \{v_1, v_2, \ldots, v_n\}$ and let X_i be the set of edges incident with v_i for $1 \le i \le n$. If $\{V_1, V_2\}$ is a cut, then observe that $\triangle_{v_i \in V_1} X_i = \triangle_{v_i \in V_2} X_i$ is the corresponding cocycle, that is, the set consisting of all cocycles of G is the subspace of $\mathscr{E}(G)$ spanned by $X_1, X_2, \ldots, X_n$, which is called the *cocycle space* $\mathscr{D}$. Since every element in $\mathcal{C}_0$ is a cocycle while each X_i is a minimal edge-cut, Lemma 1.1 implies that $\mathcal{C}_0$ and the cocycle space coincide. Note that there are $\frac{1}{2} \sum_{i=1}^{n-1} \binom{n}{i} = 2^{n-1} - 1$ cuts. When G is connected, there is a one-to-one correspondence between the cuts and nonempty cocycles and so $\dim \mathcal{C}_0 = n - 1$. In other words, $\dim \mathscr{E}(G) = |E(G)| = \dim \mathcal{C} + \dim \mathcal{C}_0$.

For a nontrivial connected graph G, observe that G is Eulerian if and only if $E(G) \in \mathcal{C}$; while G is bipartite if and only if $E(G) \in \mathcal{C}_0$. If E is a subset of $E(G)$, then $E \cap X \ne \emptyset$ for every $X \in \mathcal{C} \setminus \{\emptyset\}$ if and only if $G[E(G) \setminus E]$ is acyclic. In other words, the number of subsets E such that $E \cap X \ne \emptyset$ for every $X \in \mathcal{C} \setminus \{\emptyset\}$ equals the number of acyclic subgraphs of G. Also, $E \cap X \ne \emptyset$ for every $X \in \mathcal{C}_0 \setminus \{\emptyset\}$ if and only if $G[E]$ is a connected spanning subgraph of G. Hence, Theorem 1.10 has the following as a consequence.

Theorem 1.11. *Let G be a nontrivial connected graph.*
(a) *The number of subsets E of $E(G)$ such that $G[E]$ is contained in a spanning tree of G is odd if and only if G is Eulerian.*
(b) *The number of subsets E of $E(G)$ such that $G[E]$ contains a spanning tree of G is odd if and only if G is bipartite.*

Combining Theorems 1.3, 1.8, 1.9, and 1.11(a), we obtain a few equivalent statements.

Theorem 1.12. *For a nontrivial connected graph G, the following are equivalent:*
(a) *The graph G is Eulerian.*
(b) *The graph G is even.*
(c) *Every edge of G belongs to an odd number of cycles.*
(d) *The graph G has an odd number of cycle decompositions.*
(e) *The graph G has an odd number of acyclic subgraphs.*

Directed Eulerian Graphs

A directed circuit C in a connected digraph D is an *Eulerian circuit* if C contains every arc of D exactly once. A digraph is *Eulerian* if it contains an Eulerian circuit. The proof of the following characterization of Eulerian digraphs is essentially that of the proof of Theorem 1.3.

Theorem 1.13. *A connected digraph D is Eulerian if and only if* od v = id v *for every vertex v of D.*

A characterization of Eulerian graphs can then be given in terms of orientations of these graphs.

Theorem 1.14. *A connected graph G is Eulerian if and only if there is an Eulerian orientation of G.*

Proof. If G is an Eulerian graph of size m, then it has an Eulerian circuit $(v_1, v_2, \ldots, v_m, v_{m+1} = v_1)$. Replacing the edge $v_i v_{i+1}$ by the arc (v_i, v_{i+1}) for $1 \leq i \leq m$ produces an Eulerian orientation of G. For the converse, let D be an Eulerian orientation of a connected graph G. For every vertex $v \in V(G) = V(D)$, we have od$_D v$ = id$_D v$ by Theorem 1.13, which then implies that $\deg_G v$ = od$_D v$ + id$_D v$ is even. Thus, G is Eulerian. $\square$

There is a formula for the number of distinct Eulerian circuits in an Eulerian digraph. The theorem giving this result is called the *BEST Theorem*, obtained by two pairs of authors. This theorem is named for the initials of these four mathematicians. One pair of researchers is Nicolaas de **B**ruijn and Tatyana Pavlovna van Aardenne-**E**hrenfest [68]. The famous combinatorial concept of de Bruijn sequences was named for the Dutch mathematician de Bruijn. The father of van Aardenne-Ehrenfest was the famous physicist Paul Ehrenfest. The well-known physicist Albert Einstein was often a houseguest of the family. The other pair of mathematicians is Cedric **S**mith and William **T**utte [67]. Smith was known for his research in genetic statistics, while Tutte was one of the great graph theorists of all time.

In order to state the BEST Theorem, it is necessary to define a matrix M associated with a digraph D and some related results. Suppose that $V(D) = \{v_1, v_2, \ldots, v_n\}$. The *adjacency matrix* $A = [a_{ij}]$ of D is the $n \times n$ matrix defined by $a_{ij} = 1$ if (v_i, v_j) is an arc of D and $a_{ij} = 0$ otherwise. The *outdegree matrix* $B = [b_{ij}]$ of D is the $n \times n$ diagonal matrix for which b_{ii} = od v_i and $b_{ij} = 0$ if $i \neq j$. The matrix M is defined by $M = B - A$. For $1 \leq i, j \leq n$, the (i, j)-cofactor of M is $(-1)^{i+j} \det(M_{ij})$, where M_{ij} is the $(n-1) \times (n-1)$ submatrix of M obtained by deleting row i and column j of M and $\det(M_{ij})$ is the determinant of M_{ij}. Thus, the sum of the n entries in each row equals 0. For each i ($1 \leq i \leq n$), it is known that the values of the (i, j)-cofactors of M are the same fixed constant.

Lemma 1.2. *If X is a square matrix with the sum of the entries of each row being zero, then the (i_1, j_1)- and (i_2, j_2)-cofactors of X are equal whenever $i_1 = i_2$.*

Proof. For an $n \times n$ matrix X, let $\overrightarrow{c_i}$ be the column i vector of X. Suppose that i, j_1, and j_2 are integers satisfying $1 \le i, j_1, j_2 \le n$ and $j_1 \ne j_2$. Let Y be the $n \times n$ matrix obtained from X by replacing $\overrightarrow{c_{j_2}}$ by $-\overrightarrow{c_{j_1}}$. Thus,

$$\det(Y_{ij_1}) = (-1)^{j_1+j_2} \det(Y_{ij_2}) = (-1)^{j_1+j_2} \det(X_{ij_2}). \tag{1.6}$$

Now column j_2 of Y equals $-\overrightarrow{c_{j_1}}$, which is the sum of the columns of X except $\overrightarrow{c_{j_1}}$. Since adding a column to another does not alter the determinant, it follows that $\det(X_{ij_1}) = \det(Y_{ij_1})$. Thus, together with (1.6), we obtain $\det(X_{ij_1}) = (-1)^{j_1+j_2} \det(X_{ij_2})$. □

For a nontrivial tree T and a vertex $v \in V(T)$, the *in-tree T_v rooted at v* is the orientation of T such that there exists a directed $u - v$ path for every $u \in V(T)$. In terms of out-degree, therefore, the directed tree T_v has the property that

$$\text{od } u = \begin{cases} 0 & \text{if } u = v \\ 1 & \text{if } u \ne v. \end{cases}$$

The following is due to Tutte [66], which determines the number of spanning in-trees of a digraph rooted at a specified vertex.

Theorem 1.15 (Directed Matrix Tree Theorem). *Let D be a loopless digraph with $V(D) = \{v_1, v_2, \ldots, v_n\}$. Then the number of spanning in-trees of D rooted at v_i equals the (i, i)-cofactor of M.*

For example, let D be the digraph shown in Fig. 1.7. The matrix M for this digraph is:

$$M = \begin{bmatrix} 1 & 0 & 0 & 0 & 0 & 0 & -1 \\ 0 & 1 & -1 & 0 & 0 & 0 & 0 \\ 0 & 0 & 2 & -1 & 0 & 0 & -1 \\ 0 & 0 & -1 & 2 & -1 & 0 & 0 \\ 0 & 0 & 0 & 0 & 2 & -1 & -1 \\ -1 & 0 & 0 & 0 & -1 & 2 & 0 \\ 0 & -1 & 0 & -1 & 0 & -1 & 3 \end{bmatrix}$$

Since D is Eulerian, the out-degree and in-degree of each vertex are equal, which results in zero column sum as well as zero row sum in M. Thus, every two cofactors of M are equal by Lemma 1.2. This common cofactor is 9 and the nine spanning in-trees of D rooted at v_1 are shown in Fig. 1.8 (with v_1 colored red).

Choose one of these in-trees, say the tree T^* in Fig. 1.9, where its edges are colored red. For each vertex v_i, let s_i be an ordering of the d_i vertices adjacent from v_i in D such that if $v_i \ne v_1$, then u is the last vertex in s_i if (v_i, u) belongs to T^*.

Fig. 1.7 An Eulerian
digraph D

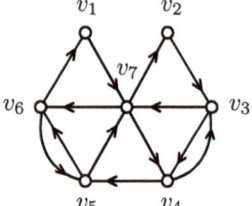

Fig. 1.8 The nine spanning
in-trees of D rooted at v_1

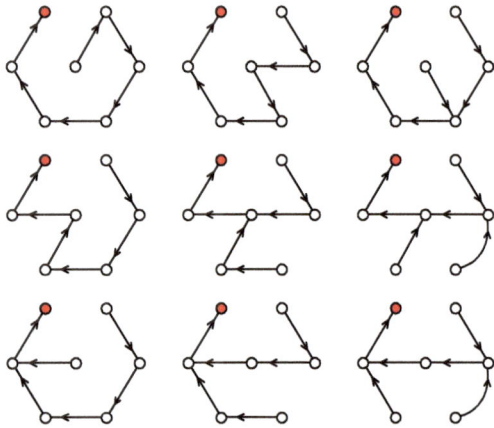

Then beginning at v_1, one can construct a unique Eulerian circuit in D by selecting
available vertices at each step as the predetermined orderings suggest. For example,
the orderings

$$s_1 : v_7 \qquad s_4 : v_3, v_5 \qquad s_6 : v_5, v_1$$
$$s_2 : v_3 \qquad s_5 : v_7, v_6 \qquad s_7 : v_2, v_6, v_4$$
$$s_3 : v_7, v_4$$

give rise to the Eulerian circuit

$$(v_1, v_7, v_2, v_3, v_7, v_6, v_5, v_7, v_4, v_3, v_4, v_5, v_6, v_1). \tag{1.7}$$

By letting $s_i' = s_i$ for $1 \leq i \leq 6$ and $s_7' : v_6, v_2, v_4$, we obtain another Eulerian
circuit

$$(v_1, v_7, v_6, v_5, v_7, v_2, v_3, v_7, v_4, v_3, v_4, v_5, v_6, v_1),$$

which is distinct from (1.7). The key observation here is that the number of distinct
Eulerian circuits generated by each in-tree in this manner equals the number of
distinct n orderings $s_1, s_2, \ldots, s_n$, which is exactly $\prod_{i=1}^{n}(d_i - 1)!$, where $d_i =$
od $v_i = $ id v_i for $1 \leq i \leq n$.

Theorem 1.16 (The BEST Theorem). *Let D be an Eulerian digraph of order n
with $V(D) = \{v_1, v_2, \ldots, v_n\}$, where* od $v_i = $ id $v_i = d_i$ *and c is the common*

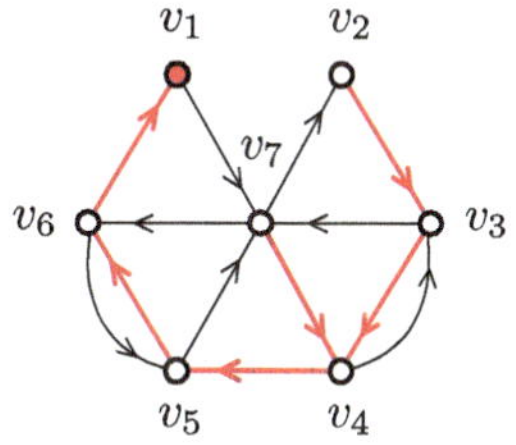

Fig. 1.9 A spanning in-tree T^* of D rooted at v_1

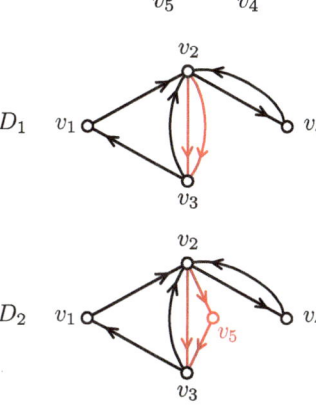

Fig. 1.10 Eulerian digraphs D_1 and D_2

cofactor of the matrix M. Then the number of distinct Eulerian circuits in D is given by $c \prod_{i=1}^{n}(d_i - 1)!$.

Therefore, the digraph D in Fig. 1.7 contains $9 \prod_{i=1}^{7}(d_i - 1)! = 18$ distinct Eulerian circuits.

Another Eulerian digraph D_1 is shown in Fig. 1.10. As this example shows, it is possible that there is more than one arc from a vertex to another. (The two arcs from v_2 to v_3 in D_1 are colored red.) In this case, consider the digraph D_2 corresponding to D_1, also shown in Fig. 1.10. One can verify that D_2 has exactly four distinct Eulerian circuits, from which one can conclude that so does D_1.

More generally, Theorem 1.16 can be applied to every Eulerian digraph (where multiple arcs and loops are allowed) if the adjacency matrix A is replaced by the square matrix A' of the same size whose (i, j)-entry equals the number of arcs from v_i to v_j. For D_1, we have

$$
M = B - A' = \begin{bmatrix} 1 & -1 & 0 & 0 \\ 0 & 3 & -2 & -1 \\ -1 & -1 & 2 & 0 \\ 0 & -1 & 0 & 1 \end{bmatrix}.
$$

The cofactor of M is 2 and there are $2 \prod_{i=1}^{4}(d_i - 1)! = 4$ distinct Eulerian circuits, as observed earlier.

1.3   Graphs with Odd Vertices

We saw in Theorem 1.4 that traversable graphs are precisely those graphs that are connected and contain exactly two odd vertices. Just as Theorem 1.14 gives a characterization of Eulerian graphs in terms of orientations, there is an analogous characterization of traversable graphs.

A directed trail T in a connected digraph D is an *Eulerian trail* if T is an open trail containing every arc of D exactly once. A characterization of digraphs having Eulerian trails is stated below.

Theorem 1.17. *A connected digraph D has an Eulerian trail if and only if D contains two vertices u and v such that $\mathrm{od}\,u = \mathrm{id}\,u + 1$ and $\mathrm{id}\,v = \mathrm{od}\,v + 1$, while $\mathrm{od}\,w = \mathrm{id}\,w$ for all vertices w of D distinct from u and v. In this case, each Eulerian trail of D is a $u - v$ trail.*

We now state the aforementioned characterization of traversable graphs.

Theorem 1.18. *A connected graph G is traversable if and only if there is an orientation of G that contains an Eulerian trail.*

If a graph G has four or more odd vertices, then G contains neither an Eulerian circuit nor an Eulerian trail, which again explains why there was no journey about Königsberg that crossed each bridge exactly once. Even though these graphs contain neither Eulerian circuits nor Eulerian trails, there are some interesting properties that they possess.

Theorem 1.19. *If G is a connected graph containing $2k \geq 2$ odd vertices, then G can be decomposed into k open trails but no fewer.*

Proof. If G is decomposed into open trails, then the vertices that are neither initial nor terminal vertices of any of these trails are even. It then follows that there is no decomposition of G into fewer than k open trails, since there are $2k$ odd vertices.

Suppose that $v_1, v_2, \ldots, v_{2k}$ are the odd vertices of G. We construct a new graph H from G by adding k new vertices in $U = \{u_1, u_2, \ldots, u_k\}$ and joining u_i to both v_i and v_{i+k} for $1 \leq i \leq k$. Then H contains an Eulerian circuit C. Deleting the k vertices in U from C results in k pairwise edge-disjoint trails in G connecting pairs of odd vertices and every edge of G lies on exactly one of these trails.   □

It is possible to have some conditions on the lengths of trails in the preceding theorem. A trail T is called an *even (odd) trail* if its length $L(T)$ is even (odd). We first present a lemma.

Lemma 1.3. *Let T_i be a trail connecting u_i and v_i in a graph G for $i = 1, 2$, where u_1, u_2, v_1, v_2 are four distinct vertices. If T_1 and T_2 are edge-disjoint and w*

Fig. 1.11 Illustrating
Lemma 1.3

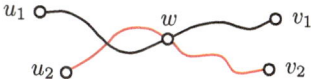

is a vertex belonging to both trails, then there are a $u_1 - u_2$ trail T_1' and a $v_1 - v_2$ trail T_2' in G that are edge-disjoint such that w belongs to both T_1' and T_2' and $E(T_1) \cup E(T_2) = E(T_1') \cup E(T_2')$. Similarly, there are a $u_1 - v_2$ trail T_1'' and a $v_1 - u_2$ trail T_2'' in G that are edge-disjoint such that w belongs to both T_1'' and T_2'' and $E(T_1) \cup E(T_2) = E(T_1'') \cup E(T_2'')$. In particular, if both T_1 and T_2 are odd trails, then either both T_1' and T_2' are even trails or both T_1'' and T_2'' are even trails.

Although Lemma 1.3 is not surprising (see Fig. 1.11, where T_1 is in black and T_2 is in red), it will be useful when proving the following result, established in [4]. The distance $d(G_1, G_2)$ between two subgraphs G_1 and G_2 in a connected graph G is given by $d(G_1, G_2,) = \min\{d(u, v) : u \in V(G_1), v \in V(G_2)\}$.

Theorem 1.20. *Let G be a connected graph containing $2k$ (≥ 2) odd vertices. Among all decompositions of G into k open trails, let s be the maximum number of such trails that are odd.*
(a) *If the size of G is even, then so is s and for every even integer s' with $0 \leq s' \leq s$, there exists a decomposition of G into k open trails, exactly s' of which are odd trails.*
(b) *If the size of G is odd, then so is s and for every odd integer s'' with $1 \leq s'' \leq s$, there exists a decomposition of G into k open trails, exactly s'' of which are odd trails.*

Proof. That $|E(G)| \equiv s \pmod 2$ is immediate. We only verify (a) as the proof of (b) is similar. Since the result is trivially true when $s = 0$, we assume that $2 \leq s \leq k$. It suffices to show that there is a decomposition of G into k open trails, exactly $s - 2$ of which are odd. Consider the decompositions of G into s odd open trails and $k - s$ even open trails. For each such decomposition $\mathscr{D}$ of G, let

$$f(\mathscr{D}) = \min\{d(T', T'') : T', T'' \in \mathscr{D} \text{ are distinct odd trails.}\}$$

and let $\mathscr{D}_0$ be among those decompositions such that $f(\mathscr{D}_0)$ is minimum. We claim that $f(\mathscr{D}_0) = 0$. Assume that $\mathscr{D}_0 = \{T_1, T_2, \ldots, T_k\}$, where each T_i ($1 \leq i \leq s$) is an odd trail. If $f(\mathscr{D}_0) > 0$, then without loss of generality, let $x \in V(T_1)$ and $y \in V(T_2)$ such that $f(\mathscr{D}_0) = d(x, y) = d(T_1, T_2)$. Let $(x = v_0, v_1, v_2, \ldots, v_d = y)$ be an $x - y$ geodesic, where then $d = d(x, y)$. Since $d(v_1, y) < f(\mathscr{D}_0)$, the trail in $\mathscr{D}_0$ that contains the edge $v_0 v_1$ must be even, say $v_0 v_1 \in E(T_{s+1})$. Also, since the initial and terminal vertices of the trails T_1 and T_{s+1} are distinct, Lemma 1.3 guarantees the existence of edge-disjoint trails T_1' and T_{s+1}', each connecting odd vertices of G, such that $E(T_1) \cup E(T_{s+1}) = E(T_1') \cup E(T_{s+1}')$. Furthermore, we may assume that T_1', say, is an odd trail and $v_0 v_1 \in E(T_1')$, while T_{s+1}' is then an even trail. Replacing

T_1 and T_{s+1} in $\mathcal{D}_0$ by T_1' and T_{s+1}', we obtain a decomposition $\mathcal{D}_1$ of G into s odd open trails and $k - s$ even open trails, where $f(\mathcal{D}_1) \leq d(T_1', T_2) \leq d(v_1, y) < f(\mathcal{D}_0)$. This is a contradiction.

Thus, $f(\mathcal{D}_0) = 0$ as claimed, that is, T_1 and T_2, say, share some common vertex. By Lemma 1.3, therefore, we can find a decomposition $\mathcal{D}_2 = \{T_1', T_2', T_3, \ldots, T_k\}$ of G into k open trails, where T_1' and T_2' are even trails. As a result, exactly $s - 2$ trails in $\mathcal{D}_2$ are odd trails. □

The following 1973 result by Chartrand, Polimeni, and Stewart [21] is a consequence of Theorem 1.20.

Theorem 1.21. *If G is a connected graph containing $2k$ (≥ 2) odd vertices, then G can be decomposed into k open trails, at most one of which is an odd trail.*

If G is a connected graph of even size with $2k$ (≥ 4) odd vertices, then it may not be possible to decompose G into k open trails where some of them have odd length. Consider, for example, the star $K_{1,2k}$. It can be shown that if G is a connected bipartite graph of even size with four odd vertices, then G can be decomposed into two odd open trails if and only if each partite set of G contains exactly two odd vertices.

Problem 1.1. Suppose that G is a nontrivial connected graph of even size.
(a) If G contains exactly four odd vertices and G is not bipartite, then under what conditions can G be decomposed into two odd open trails?
(b) If G contains $2k$ (≥ 2) odd vertices, then under what conditions can G be decomposed into open trails, at least two of which are odd trails?

The fact that a connected graph G containing $2k$ odd vertices ($k \geq 1$) can be decomposed into k open trails connecting pairs of odd vertices implies that G has k pairwise edge-disjoint paths connecting pairs of odd vertices. We cannot always specify which pairs have this property, however. Should G contain exactly four odd vertices, every two pairs of these four odd vertices are connected by edge disjoint paths.

Theorem 1.22. *Suppose that G is a connected graph containing exactly four odd vertices. If $\{u_1, v_1\}$ and $\{u_2, v_2\}$ are disjoint 2-subsets of the set of odd vertices of G, then there exist paths $P^{(i)}$ ($i = 1, 2$) in G connecting u_i and v_i that are edge-disjoint.*

Proof. It suffices to show that there are a $u_1 - v_1$ trail and a $u_2 - v_2$ trail that are edge-disjoint. By Theorem 1.19, the graph G can be decomposed into two open trails T_1 and T_2, where say T_i is an $x_i - y_i$ trail for $i = 1, 2$. Thus, $\{u_1, u_2, v_1, v_2\} = \{x_1, x_2, y_1, y_2\}$ is the set of odd vertices in G. Since the result is immediate if $\{u_1, v_1\} \in \{\{x_1, y_1\}, \{x_2, y_2\}\}$, we may assume, for example, that $\{u_1, v_1\} = \{x_1, y_2\}$. Since G is connected, let z be a vertex belonging to both T_1

Fig. 1.12 A graph
containing four odd vertices

G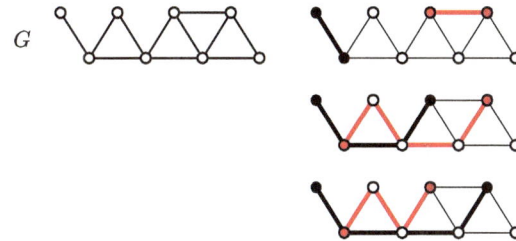

and T_2. For $i = 1, 2$, let T_i' and T_i'' be an $x_i - z$ subtrail and $z - y_i$ subtrail of T_i, respectively, such that $E(T_i') \cap E(T_i'') = \emptyset$. Then T_1' followed by T_2'' is an $x_1 - y_2$ trail while T_2' followed by T_1'' is an $x_2 - y_1$ trail and these trails are edge-disjoint.   □

The graph G in Fig. 1.12 contains exactly four odd vertices. Paris of edge-disjoint paths connecting the odd vertices are shown in bold black and red lines.

Theorem 1.22 does not necessarily hold if there are more than four odd vertices. For example, suppose that a connected graph G contains a bridge e. Then the number of odd vertices belonging to V_i must be odd for $i = 1, 2$, where $\{V_1, V_2\}$ is the partition of $V(G)$ such that each V_i induces a component in $G - e$. Thus, if G has the property that for every four distinct odd vertices u_1, u_2, v_1, v_2 of G, there are $u_i - v_i$ paths $P^{(i)}$ in G ($i = 1, 2$) that are edge-disjoint, then one of the sets V_1 and V_2 must contain exactly one odd vertex. However, this is not necessarily the case if there are more than four odd vertices. Consider, for example, the double star (a tree whose diameter equals 3) of order at least 6 in which each of the two central vertices is adjacent to at least two end-vertices.

Problem 1.2. Let G be a connected graph containing $2k$ (≥ 6) odd vertices. If G contains a bridge e, then let $\{V_1, V_2\}$ be the partition of $V(G)$ such that each V_i induces a component in $G - e$. Furthermore, suppose that $\min\{|V_i \cap V_{odd}| : i = 1, 2\} = 1$, where V_{odd} is the set of odd vertices in G. If $\{\{u_1, v_1\}, \{u_2, v_2\}, \ldots, \{u_k, v_k\}\}$ is a partition of V_{odd}, then is it possible to find $u_i - v_i$ paths $P^{(i)}$ ($i = 1, 2, \ldots, k$) that are pairwise edge-disjoint? What if G contains no bridge?

1.4   The Chinese Postman Problem

In 1962 the Chinese mathematician Meigu Guan (often known as Mei-Ko Kwan) introduced a problem that has a connection with Eulerian graphs.

Suppose that a postman starts from the post office and has mail to deliver to the houses along each street of his mail route. Once he has completed delivering the mail, he returns to the post office. The problem is to find the minimum length of a round trip that accomplishes this. Alan Goldman coined a name for this problem by which it is commonly known.

The Chinese Postman Problem. *Determine the minimum length of a round trip that traverses every road in a mail route at least once.*

The Chinese Postman Problem suggests a problem in graph theory. Let G be a connected graph. What is the minimum length ℓ of a closed walk that contains every edge of G at least once? Of course, if G is an Eulerian graph, then this minimum length ℓ equals the size m of G. Suppose, on the other hand, that G is not Eulerian. Then $\ell > m$. Let H be the multigraph obtained from G by replacing each edge uv of G by two parallel edges joining u and v. Then H is Eulerian and thus H contains an Eulerian circuit which corresponds to a closed walk in G that encounters every edge of G exactly twice. This observation was made by Euler in paragraph 18 of his 1736 Königsberg paper and implies that every connected graph G of size m has a closed walk that contains every edge of G at least once and at most twice. In particular, the minimum length ℓ of a closed walk containing every edge of G satisfies $m \leq \ell \leq 2m$, at least. As we mentioned, $\ell = m$ if and only if G is Eulerian. So we may assume that G contains $2k$ odd vertices where $k \geq 1$. There are

$$(2k - 1)(2k - 3)(2k - 5) \cdots 3 \cdot 1 = \frac{(2k - 1)!}{2^{k-1}(k - 1)!}$$

ways to divide these $2k$ odd vertices into k pairs. For each of these k pairs $\{u, v\}$, we compute $d(u, v)$ and sum these distances over all k pairs. We then compute this number for all such sets of pairs. Suppose that the minimum of these numbers is d. Then $\ell = m + d$. If the k pairs that produce the number d are $\{u_1, v_1\}$, $\{u_2, v_2\}$, ..., $\{u_k, v_k\}$, then $\sum_{i=1}^{k} d(u_i, v_i) = d$. If $P^{(i)}$ is a $u_i - v_i$ geodesic in G, then these paths $P^{(i)}$ $(1 \leq i \leq k)$ are pairwise edge disjoint. If we were to replace each edge uv that lies on any of these paths by two parallel edges joining u and v, then the resulting multigraph H is Eulerian, which implies that there is a closed walk in G that contains every edge twice that belongs to one of the paths $P^{(i)}$ and contains each of all other edges once.

As a consequence of the observation above, for every connected graph G, there exists a closed walk containing every edge of G the same number of times. In fact, there exists a closed walk containing every edge of G exactly k times if k is a positive even integer. For a positive odd integer k, there is a closed walk containing every edge of G exactly k times if and only if G is Eulerian. This brings up the question of which connected graphs have a closed walk in which no two edges appear on the walk the same number of times. For the purpose of this discussion, it is useful to introduce some additional terminology.

A closed walk containing every edge of a connected graph at least once is called an *Eulerian walk*. By a *minimum Eulerian walk*, we mean an Eulerian walk of minimum length. Thus the length $e(G)$ of a minimum Eulerian walk in a connected graph G of size m is at least m and at most $2m$. Furthermore, this number is m if and only if G is Eulerian and this number is $2m$ if and only if G is a tree. In order to show that the length of a minimum Eulerian walk in a connected graph G of size m is exactly $2m$ if and only if G is a tree, we state a theorem due to Kwan [48].

Theorem 1.23 (Kwan's Theorem). *A minimum Eulerian walk W in a connected graph G encounters no edge of G more than twice. Also, no more than half of the edges in any cycle appear twice in W.*

Theorem 1.24. *Let G be a nontrivial connected graph. Then $e(G) = 2|E(G)|$ if and only if G is a tree.*

Proof. By Kwan's Theorem, it suffices to show that $e(G) \geq 2|E(G)|$ if G is a tree.

If uv is a bridge in a connected graph G, then consider an Eulerian walk W of G with initial vertex u. The first time that v is encountered on W, it is preceded by u; while the next time that u is encountered on W, it is preceded by v. Therefore, the edge uv occurs an even number of times on W. Since every edge is a bridge in a tree, the length $L(W)$ of W is at least twice of the size of G when G is a tree. □

As we mentioned before, while the Chinese Postman Problem asks for the minimum length of a closed walk in a connected graph G such that every edge of G appears on the walk once or twice, another problem of interest is that of determining the minimum length of a closed walk in G such that no two edges of G appear the same number of times in the walk. Such walks in a graph G distinguish the edges of G by their occurrences on the walk, which gives rise to the concept of irregular Eulerian walks in graphs [4].

An Eulerian walk W in a connected graph G is called *irregular* if no two edges of G occur the same number of times in W. The first observation in this connection is that every connected graph contains an irregular Eulerian walk. Suppose that $E(G) = \{e_1, e_2, \ldots, e_m\}$, where $e_i = u_i v_i$ for $1 \leq i \leq m$. If we replace e_i by $2i$ parallel edges joining u_i and v_i for $i = 1, 2, \ldots, m$, then the resulting multigraph H is Eulerian, which implies that G has an irregular Eulerian walk W that contains the edge e_i a total of $2i$ times. The length ℓ of a minimum irregular Eulerian walk therefore satisfies the following:

$$\binom{m+1}{2} \leq \ell \leq 2\binom{m+1}{2}. \tag{1.8}$$

The upper bound in (1.8) is attained only when a connected graph is a tree, as we show next.

Theorem 1.25. *For a nontrivial connected graph G of size m, the length of a minimum irregular Eulerian walk is $2\binom{m+1}{2}$ if and only if G is a tree.*

Proof. Let W be an irregular Eulerian walk in G. As we saw in the proof of Theorem 1.24, each bridge in G is encountered an even number of times on W. Thus, if G is a tree, then each edge of G is encountered an even number of times on W, that is, $L(W) \geq 2\binom{m+1}{2}$. It then follows by (1.8) that the length of a minimum irregular Eulerian walk is exactly $2\binom{m+1}{2}$.

If G is not a tree, then G contains some cycles. By Kwan's theorem, there is an Eulerian walk W_1 in which no edge of G occurs on W_1 more than twice and some edges occur on W_1 exactly once. Let $e_1, e_2, \ldots, e_k$ be those edges occurring exactly once on W_1. Also, let $e'_1, e'_2, \ldots, e'_\ell$ be those edges occurring exactly twice on W_1. Replacing the edge $e_i = u_i v_i$ by $2i - 1$ parallel edges joining u_i and v_i $(1 \le i \le k)$ and replacing the edge $e'_i = u'_i v'_i$ by $2i$ parallel edges joining u'_i and v'_i $(1 \le i \le \ell)$, we obtain an Eulerian multigraph M. This implies that there is an Eulerian walk W_2 in G where e_i appears $2i - 1$ times while e'_i appears $2i$ times. Since G contains cycles, $k \ge 1$ and so $L(W_2) < 2\binom{m+1}{2}$. Hence, the length of a minimum irregular Eulerian walk is also less than $2\binom{m+1}{2}$. $\square$

If a graph G of size m contains an irregular Eulerian walk W of length $\binom{m+1}{2}$, then the walk W is said to be *optimal*. Necessarily, every optimal irregular Eulerian walk is a minimum irregular Eulerian walk. The following result characterizes those graphs containing an optimal irregular Eulerian walk.

Theorem 1.26. *A connected graph G of size m contains an optimal irregular Eulerian walk if and only if G contains an even subgraph of size $\lceil m/2 \rceil$.*

Proof. First, assume that G contains a subgraph H of size $\lceil m/2 \rceil$ such that $\deg_H v$ is even for every $v \in V(H)$. Let $E(H) = \{e_1, e_2, \ldots, e_{\lceil m/2 \rceil}\}$ and $E(G) = E(H) \cup \{e'_1, e'_2, \ldots, e'_{\lfloor m/2 \rfloor}\}$. Replacing (i) each edge e_i by $2i - 1$ parallel edges and (ii) each edge e'_i by $2i$ parallel edges, we obtain an Eulerian multigraph M. Then an Eulerian circuit in M gives rise to an irregular Eulerian walk W in G in which e_i appears exactly $2i - 1$ times $(1 \le i \le \lceil m/2 \rceil)$ and e'_i appears exactly $2i$ times $(1 \le i \le \lfloor m/2 \rfloor)$. Then $L(W) = \binom{m+1}{2}$ and so W is an optimal irregular Eulerian walk in G.

For the converse, suppose that G contains an optimal irregular Eulerian walk W. We may assume that $E(G) = \{e_1, e_2, \ldots, e_m\}$, where e_i appears exactly i times on W. Let $E_1 = \{e_1, e_3, \ldots, e_{2\lfloor (m-1)/2 \rfloor +1}\}$ and $E_2 = E(G) \setminus E_1$. If H_1 and H_2 are the subgraphs induced by E_1 and E_2, respectively, then $\{H_1, H_2\}$ is a decomposition of G. We claim that the vertices in H_1 are all even. The existence of W guarantees that the multigraph M obtained from G by replacing each edge e_i by i parallel edges is Eulerian. For $j = 1, 2$, let M_j be the multigraph obtained by replacing each edge $e_i \in E(H_j)$ by i parallel edges. Then $\{M_1, M_2\}$ is a decomposition of M. If $v \in V(H_1)$, then $\deg_{M_1} v = \deg_M v - \deg_{M_2} v$ is even since both $\deg_M v$ and $\deg_{M_2} v$ are. Therefore, $\deg_{H_1} v$ must be even since $\deg_{M_1} v$ equals the sum of $\deg_{H_1} v$ odd integers. $\square$

By Theorem 1.26, neither K_2 nor K_3 contains optimal irregular Eulerian walks. For $4 \le n \le 6$, on the other hand, $C_3 \subseteq K_4$, $C_5 \subseteq K_5$, and $K_{2,4} \subseteq K_6$ and so K_n has an optimal irregular Eulerian walk. In general, for an integer $n \ge 4$, let $M = \binom{n}{2}$ if n is odd and $M = n(n-2)/2$ if n is even. By recalling Alspach's conjecture (that is no longer a conjecture), which basically lists possible cycle decompositions of complete graphs of odd order and $K_{2,2,\ldots,2}$, we see that if $3 \leq m \leq M - 3$ or $m = M$, then K_n contains an even subgraph H of size m. For example, for each integer $n \geq 7$, Alspach's conjecture guarantees that there is a decomposition $\mathscr{D}_1 \cup \mathscr{D}_2$ of K_n, where

$$
\mathscr{D}_1 = \begin{cases}
\{ (n/4)C_{n-1} \} & \text{if } n \equiv 0 \pmod 4 \\
\{ \lfloor n/4 \rfloor C_n \} & \text{if } n \equiv 1 \pmod 4 \\
\{ (n/2 - 1)C_{n/2}, C_{\lceil n/4 \rceil} \} & \text{if } n \equiv 2 \pmod 4 \\
\{ (\lfloor n/2 \rfloor - 1)C_{\lfloor n/2 \rfloor}, C_{\lfloor 3n/4 \rfloor} \} & \text{if } n \equiv 3 \pmod 4
\end{cases}
$$

$$
\mathscr{D}_2 = \begin{cases}
\{ (n/4)C_{n-3}, (n/2)K_2 \} & \text{if } n \equiv 0 \pmod 4 \\
\{ \lfloor n/4 \rfloor C_n \} & \text{if } n \equiv 1 \pmod 4 \\
\{ (n/2 - 2)C_{n/2-1}, C_{\lfloor 3n/4 \rfloor - 2}, (n/2)K_2 \} & \text{if } n \equiv 2 \pmod 4 \\
\{ (\lfloor n/2 \rfloor - 1)C_{\lfloor n/2 \rfloor}, C_{\lfloor 3n/4 \rfloor - 1} \} & \text{if } n \equiv 3 \pmod 4.
\end{cases}
$$

By combining the cycles in $\mathscr{D}_1$, we obtain an even sugraph $H \subseteq K_n$ whose size equals $\lceil \binom{n}{2}/2 \rceil$.

Theorem 1.27. *The complete graph of order n contains an optimal irregular Eulerian walk if and only if $n \geq 4$.*

For the complete bipartite graphs having optimal irregular Eulerian walks, we have the following characterization.

Theorem 1.28. *For integers n_1 and n_2 with $2 \leq n_1 \leq n_2$, the complete bipartite graph K_{n_1,n_2} has an optimal irregular Eulerian walk if and only if either (i) n_1 and n_2 are both even and either $n_1 \neq 2$ or $4 \mid n_2$ or (ii) at least one of n_1 and n_2 is odd and $n_1 n_2 \equiv 0, 3 \pmod 4$.*

1.5   Randomly Eulerian Graphs

Some Eulerian graphs G contain vertices v with a rather unusual property, namely every trail T in G with initial vertex v can be extended to an Eulerian circuit. Graphs with this property have been referred to by many names but we will say that these graphs are *randomly Eulerian from v*. For example, each of the Eulerian graphs G_i ($1 \leq i \leq 4$) in Fig. 1.13 is randomly Eulerian from a vertex v if and only if v is colored red. Thus, the graph G_1 is randomly Eulerian from no vertex, G_2 and G_3 are randomly Eulerian from some but not every vertex, while G_4 is randomly Eulerian from every vertex.

Ore [57, pp. 74–76] characterized those graphs that are randomly Eulerian from a vertex.

Fig. 1.13 Graphs that are randomly Eulerian from some vertex

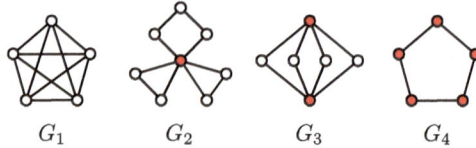

$$G_1 \qquad\qquad G_2 \qquad\qquad G_3 \qquad\qquad G_4$$

Theorem 1.29. *An Eulerian graph G is randomly Eulerian from a vertex v if and only if every cycle in G contains v.*

Proof. First, suppose that a graph G is randomly Eulerian from a vertex v but has a cycle C that does not contain v. If H is the graph obtained from G by deleting the edges of C, then every nontrivial component of H is Eulerian and so has a cycle decomposition. In particular, the component H_1 containing v is nontrivial since $\deg_{H_1} v = \deg_G v \geq 2$. Then find a cycle decomposition $\mathscr{D}_1$ of H_1 and construct a circuit C' beginning and ending at v by successively traversing the cycles in $\mathscr{D}_1$ that contain v. Therefore, C' is a circuit in G that contains no edges of C but cannot be extended any further, since every edge incident with v is already in C'. This is a contradiction.

We now verify the converse. Let G be an Eulerian graph with the property that every cycle of G contains a vertex v of G. Let T be a trail in G with initial vertex v such that it cannot be extended to a longer trail. Necessarily, T is a circuit that contains all edges that are incident with v in G. Now let H be the graph obtained from G by deleting the edges in T. If H contains a nontrivial component H', then H' is an Eulerian graph and so contains cycles to which v does not belong. Since this contradicts the assumption that every cycle in G contains v, it follows that $E(T) = E(G)$, that is, T is an Eulerian circuit. $\qquad\square$

As a consequence of Ore's and Veblen's theorems (Theorems 1.29 and 1.5), we have the following.

Theorem 1.30. *An Eulerian graph G is randomly Eulerian from a vertex v if and only if every cycle in a cycle decomposition of G contains v and no two cycles in a decomposition have more than one other vertex in common.*

Therefore, an Eulerian graph G is randomly Eulerian from every vertex if and only if G is a cycle. For $n \geq 5$, let G be a graph of order n that is randomly Eulerian from a vertex v. By Theorem 1.29, the graph $G - v$ is acyclic. Furthermore, if $G - v$ is disconnected or contains at least two vertices of degree greater than 2, then v is the only vertex belonging to every cycle in G. Otherwise, $G - v$ is a star or a subdivision of a star (including paths). In particular, when $G - v$ is (a subdivision of) a star but not a path, then G is (a subdivision of) $K_{1,1,2\ell+1}$ ($\ell \geq 1$) in which v and another vertex, say u, have degree $2(\ell + 1)$ while each of the remaining vertices has degree 2. In this case, G is randomly Eulerian from both u and v but from no

Fig. 1.14 Graphs that are randomly traversable from two, one, or no vertices

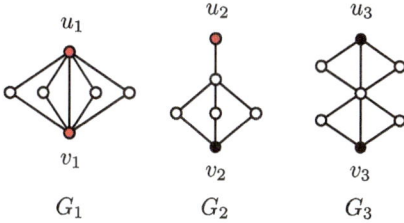

G_1 G_2 G_3

other vertices. We therefore have the following result, which shows that the graphs in Fig. 1.13 illustrate all possible situations regarding randomly Eulerian graphs.

Theorem 1.31. *Every Eulerian graph G is randomly Eulerian from no, one, two, or all vertices of G.*

There is an analogous concept for traversable graphs. Let G be a traversable graph in which u and v are the odd vertices. Then G is said to be *randomly traversable from u* if every trail with initial vertex u can be extended to an Eulerian trail T in G. Necessarily then, T is a $u - v$ trail. The following characterization of such graphs was given by Chartrand and White [18] in 1970. The same result was then independently obtained by Dirac [30] in 1972. We present a slightly different proof by making a use of Ore's theorem (Theorem 1.29).

Theorem 1.32. *A traversable graph G whose odd vertices are u and v is randomly traversable from u if and only if every cycle of G contains v.*

Proof. Let H_1 be the graph obtained from G by adding a new vertex w and the edges uw and vw. If every cycle in G contains v, then H_1 is randomly Eulerian from v by Ore's theorem. Thus, the path (v, w, u) can always be extended to an Eulerian circuit of H_1. In other words, every trail beginning at u can be extended to an Eulerian trail of G.

Conversely, if G is randomly traversable from u but has a cycle C that does not contain v, then the graph H_2 obtained from G by deleting the edges in C as well has only two odd vertices, namely u and v. Thus, H_2 has a nontrivial component H' containing u, v and all edges incident with v. However then, although H' is traversable, any $u - v$ Eulerian trail of H' cannot be extended to an Eulerian trail of G, which contradicts the fact that G is randomly traversable from u. □

Consider the three graphs G_1, G_2, and G_3 in Fig. 1.14. Note that each graph G_i is traversable and contains an Eulerian $u_i - v_i$ trail. By Theorem 1.32, G_1 is randomly traversable from both u_1 and v_1, G_2 is randomly traversable from u_2 but not from v_2, while G_3 is randomly traversable from neither u_3 nor v_3.

Hamiltonian Walks

2

2.1   The Icosian Game

The year 1857 saw the introduction of a two-person game called the *Icosian Game*. In this game, one player is to place some of 20 given pieces on the points of a playing board (in the shape of dodecahedron) as shown in Fig. 2.1 so that successive pieces are placed along the lines of the board. These pieces may be required to fulfill other conditions as well. The other player then has the responsibility to place the remaining pieces on the remaining points in such a way that every consecutive pair of pieces lie along a line of the board and that the twentieth piece lies along a line of the first piece. Sometimes this can be done, sometimes it cannot.

This game was the invention of William Rowan Hamilton. Before discussing this game in more detail, let us go back in history to learn some facts about Hamilton.

Hamilton was born in Dublin, Ireland in 1805. He was brought up by his uncle, who educated him. Very early on, it became clear that Hamilton was an extraordinarily talented individual. Indeed, at age 5, young William had mastered the languages Latin, Greek, and Hebrew. By the time he reached 12 years of age, he had become quite accomplished with mental arithmetic. At age 15, he had studied the work of Sir Isaac Newton and Pierre-Simon Laplace. That Hamilton discovered an error in Laplace's work on celestial mechanics brought him to the attention of the Astronomer Royal of Dublin. At the age of 18, Hamilton became a student at Trinity College Dublin. There he placed first in every examination in every subject. During his first year he was awarded "optime" in classics, which had not been awarded in 20 years. Later he was awarded "optime" in physics, an unheard of distinction to receive two such awards in different subjects. His education stopped at age 21 when he became Professor of Astronomy at Trinity College. With this came the title of Royal Astronomer of Ireland.

In 1832 Hamilton predicted that a ray of light passing through a biaxial crystal would be refracted into the shape of a cone. When this was experimentally confirmed, this resulted in a major scientific announcement. Hamilton was knighted for his discovery in 1835, thereby becoming *Sir* William Rowan Hamilton.

F. Fujie and P. Zhang, *Covering Walks in Graphs*, SpringerBriefs in Mathematics,
DOI 10.1007/978-1-4939-0305-4_2, © Futaba Fujie, Ping Zhang 2014

Fig. 2.1 The Icosian Game

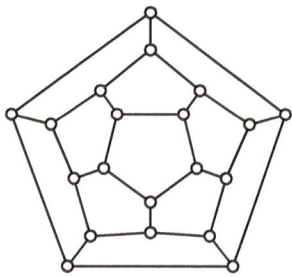

In 1835 Hamilton observed that complex numbers could be represented as ordered pairs of real numbers. For the next 8 years, Hamilton attempted to extend his theory to ordered triples but he was never successful. In 1843, however, while walking across the Brougham Bridge on the Royal Canal in Dublin, Hamilton discovered a set of 4-dimensional numbers $a + bi + cj + dk$, where $a, b, c,$ and d are real numbers, called *quaternions*. This was the first example of non-commutative algebra. He was so excited that he carved the formula he had discovered into the bridge. Today, on a plaque attached to the bridge, the following is written:

> *Here as he walked by*
> *on the 16th of October 1843*
> *Sir William Rowan Hamilton*
> *in a flash of genius discovered*
> *the fundamental formula for*
> *quaternion multiplication*
> $$\mathbf{i}^2 = \mathbf{j}^2 = \mathbf{k}^2 = \mathbf{ijk} = -1.$$

As a consequence of doing this, Hamilton had essentially introduced the cross product and dot product for vector algebra.

In 1856 Hamilton discovered a non-commutative algebraic structure referred to as the *Icosian Calculus*. This discovery came from his failed attempts to find an algebra of ordered triples that would reflect the three Cartesian axes in the Euclidean 3-space just as ordered pairs reflect the two Cartesian axes in the Euclidean plane. The symbols Hamilton used in his Icosian Calculus represented moves between vertices on a dodecahedron. This led to Hamilton's invention of the Icosian Game, which he used as a means of illustrating and popularizing his mathematical discovery. The Icosian Game was introduced to the public in 1857 at a meeting of the British Association in Dublin.

How Hamilton thought of connecting his Icosian Calculus to traveling along the edges of a dodecahedron is unknown. The mathematician John Graves was one of Hamilton's best friends. In 1859 a friend of Graves manufactured a version of the Icosian Game in the form of a small table consisting of a game board with legs, which was sent to Hamilton. Graves put Hamilton in contact with John Jaques,

Fig. 2.2 The traveler version
of the Icosian Game

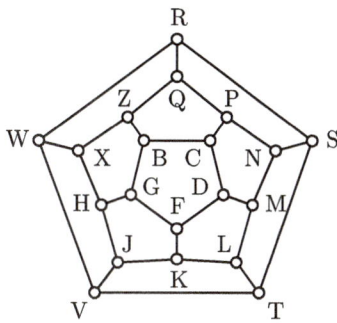

whose company John Jaques and Son manufactured toys and games. Hamilton sold
the rights to his game for 25 pounds to this manufacturer, which was later known as
John Jaques of London and then Jaques of London. This company, still in existence
after two centuries, is known for the chess sets it sells. It also invented the game of
ping pong.

Two versions of Hamilton's game were manufactured by Jaques, one played on
a flat board and another for a "traveler," played on an actual dodecahedron. The
traveler version of the game was labeled as:

NEW PUZZLE

TRAVELLER'S DODECAHEDRON

or

A VOYAGE ROUND THE WORLD

Here the 20 vertices of the dodecahedron are labeled with the 20 consonants of the
English alphabet (see Fig. 2.2), which stands for the following 20 cities:

B.	Brussels	N.	Naples
C.	Canton	P.	Paris
D.	Delhi	Q.	Quebec
F.	Frankfort	R.	Rome
G.	Geneva	S.	Stockholm
H.	Hanover	T.	Toholsk
J.	Jeddo	V.	Vienna
K.	Kashmere	W.	Washington
L.	London	X.	Xenia
M.	Moscow	Z.	Zanzibar

The goal of this game was to construct a round trip in which each of the 20 cities
would be visited exactly once. Hamilton played a role in marketing the game. The
preface to the instruction pamphlet, written by Hamilton, began as follows:

In this new Game (invented by

Sir WILLIAM ROWAN HAMILTON, LL.D., & c., of Dublin,

and by him named Icosian from a Greek word signifying 'twenty') a player is to place the whole or part of a set of twenty numbered pieces or men upon the points or in the holes of a board, represented by the diagram above drawn, in such a manner as always to proceed along the lines of the figure, and also to fulfill certain other conditions, which may in various ways be assigned by another player. Ingenuity and skill may thus be exercised in proposing as well as in resolving problems of the game. For example, the first of the two players may place the first five pieces in any five consecutive holes, and then require the second player to place the remaining fifteen men consecutively in such a manner that the succession may be cyclical, that is, so that No. 20 may be adjacent to No. 1; and it is always possible to answer any question of this kind. Thus, if B C D F G are the five given initial points, it is allowed to complete the succession by following the alphabetical order of the twenty consonants, as suggested by the diagram itself; but after placing the piece No. 6 in hole H, as above, it is also allowed (by the supposed conditions) to put No. 7 in X instead of J, and then to conclude with the succession, W R S T V J K L M N P Q Z. Other examples of Icosian Problems, with solutions of some of them, will be found in the following page.

2.2   Hamiltonian Graphs

Hamilton's Icosian Game gave rise to a much-studied class of graphs named for him. A path containing all the vertices of a graph G is a *Hamiltonian path* in G, while a cycle containing all the vertices of G is a *Hamiltonian cycle* in G. If G contains a Hamiltonian path, then G is connected; if G contains a Hamiltonian cycle, then G is 2-connected (that is, $G - v$ is connected for every $v \in V(G)$). A graph is *Hamiltonian* if it contains a Hamiltonian cycle and a graph is *traceable* if it contains a Hamiltonian path. While the graph of the dodecahedron (shown in Figs. 2.1 and 2.2) and the graph G_1 of Fig. 2.3 are both Hamiltonian, none of the graphs G_2, G_3, and G_4 of Fig. 2.3 are Hamiltonian. The graphs G_2 and G_3 have Hamiltonian paths, however. The graph G_4 does not contain a Hamiltonian path. Therefore, G_1, G_2, and G_3 are traceable but G_4 is not.

In 1855, the year before Hamilton invented the Icosian Game, Thomas Penyngton Kirkman studied questions as to whether all vertices of a polyhedron could be visited exactly once by moving along edges of a polyhedron and returning to the starting vertex. So even though Kirkman had considered this concept before Hamilton did, these paths, cycles, and graphs were named for Hamilton, not Kirkman.

Determining conditions under which a graph is Hamiltonian did not occur until 1952 when Gabriel Andrew Dirac [29] introduced a sufficient condition for a graph to be Hamiltonian in terms of the degrees of the vertices of a graph. Gabriel Dirac was the stepson of Paul Adrien Maurice Dirac, who was awarded a Nobel Prize in Physics in 1933. The smallest and largest degrees among the vertices of a graph G are the *minimum degree $\delta(G)$* and *maximum degree $\Delta(G)$*, respectively, of G.

Fig. 2.3 Illustrating
Hamiltonian and traceable
graphs

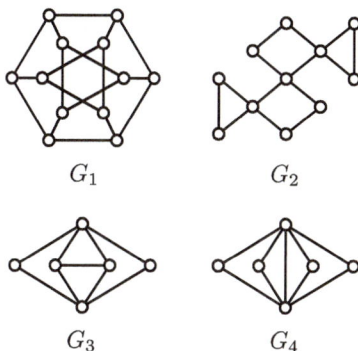

G_1 G_2

G_3 G_4

Theorem 2.1 (Dirac's Theorem). *If G is a graph of order $n \geq 3$ and $\delta(G) \geq n/2$, then G is Hamiltonian.*

Proof. Assume that this statement is false. Then for some integer $n \geq 3$, there is a non-Hamiltonian graph G of order n and maximum size for which $\deg v \geq n/2$ for each vertex v of G. Surely G is not complete, so G contains pairs of nonadjacent vertices. Let u, v be such a pair. Thus $G + uv$ is Hamiltonian and every Hamiltonian cycle of $G + uv$ necessarily contains the edge uv. This in turn implies that G contains a Hamiltonian $u - v$ path $(u = v_1, v_2, \ldots, v_n = v)$. Let $v_{n+1} = v_1$ and define $A = \{v_i : uv_{i+1} \in E(G)\}$ and $B = \{v_i : vv_i \in E(G)\}$. Since $v_n \notin A \cup B$, it follows that $|A \cup B| \leq n - 1$. Corresponding to each vertex adjacent to u is an element of A; that is, $|A| = \deg u$. Similarly, $|B| = \deg v$. Thus, $|A| + |B| \geq n$.

If there exists a vertex v_{i^*} belonging to $A \cap B$, then $2 \leq i^* \leq n - 2$ and $uv_{i^*+1}, vv_{i^*} \in E(G)$. However then,

$$(u, v_{i^*+1}, v_{i^*+2}, \ldots, v_n = v, v_{i^*}, v_{i^*-1}, \ldots, v_2, v_1 = u)$$

is a Hamiltonian cycle in G, producing a contradiction. Thus $A \cap B = \emptyset$ and so $n \leq |A| + |B| = |A \cup B| \leq n - 1$, which is clearly impossible.   □

In 1960 Oystein Ore [56] generalized Dirac's theorem. In fact, the proof of Theorem 2.1 given above also serves as a proof of Ore's theorem.

Theorem 2.2 (Ore's Theorem). *If G is a graph of order $n \geq 3$ such that $\deg u + \deg v \geq n$ for every pair u, v of nonadjacent vertices of G, then G is Hamiltonian.*

Suppose that G is a nontrivial graph and consider the graph $H = G \vee K_1$, the join of G and a vertex. This new graph H is certainly Hamiltonian if H satisfies the condition in either Theorems 2.1 or 2.2. Since G is traceable if and only if H is Hamiltonian, we obtain the following as a corollary.

Theorem 2.3. *Let G be a graph of order n $\geq$ 2.*
(a) *If $\delta(G) \geq (n-1)/2$, then G is traceable.*
(b) *If $\deg u + \deg v \geq n - 1$ for every pair u, v of nonadjacent vertices of G, then G is traceable.*

Following the publication of Ore's theorem was a succession of new sufficient conditions for a graph G to be Hamiltonian in terms of the degrees of the vertices of G, each more general than those that preceded it. The most general of these was based on Ore's theorem and is due to J. Adrian Bondy and Vašek Chvátal.

Let G be a graph of order n. If u_1 and v_1 are two nonadjacent vertices such that $\deg_G u_1 + \deg_G v_1 \geq n$, then join u_1 and v_1 by an edge producing the graph $G_1 = G + u_1v_1$. If, in G_1, there are two nonadjacent vertices u_2 and v_2 such that $\deg_{G_1} u_2 + \deg_{G_1} v_2 \geq n$, then join u_2 and v_2 by an edge producing the graph $G_2 = G_1 + u_2v_2$. This procedure is continued until no such pairs of nonadjacent vertices remain. This final graph is called the *closure* of G and is denoted by $CL(G)$. Adding the edges described above can occur in many different orders but the resulting graph is always the same graph, namely $CL(G)$. The primary importance of this concept lies in the following theorem [12].

Theorem 2.4 (Bondy and Chvátal's Theorem). *A graph is Hamiltonian if and only if its closure is Hamiltonian.*

Proof. First, if a graph G is Hamiltonian, then surely $CL(G)$ is Hamiltonian. Suppose then that G is a graph of order $n \geq 3$ such that $CL(G)$ is Hamiltonian. Let $G, G_1, G_2, \ldots, G_{k-1}, G_k = CL(G)$ be a sequence of graphs produced during the process of obtaining $CL(G)$. In particular, $CL(G) = G_k = G_{k-1} + uv$, where u and v are nonadjacent vertices in G_{k-1} and $\deg_{G_{k-1}} u + \deg_{G_{k-1}} v \geq n$. Therefore, according to the proof of Theorem 2.1, G_{k-1} is Hamiltonian. Proceeding backwards, we see that $G_{k-2}, G_{k-3}, \ldots, G_2, G_1$ and finally G are Hamiltonian. $\qquad\square$

By Theorem 2.4, a graph G of order at least 3 is Hamiltonian if $CL(G)$ is complete. A graph can be Hamiltonian without its closure being complete, however. For example, for the n-cycles C_n, which are clearly Hamiltonian, $CL(C_n) = C_n \neq K_n$ for $n \geq 5$.

The sufficient conditions presented for a graph to be Hamiltonian in Dirac's and Ore's theorems are just that, namely, they are sufficient only. That is, a graph can be Hamiltonian without satisfying either of these conditions. The fact that the cycles of order at least 5 are Hamiltonian is not a consequence of any of the theorems by Dirac, Ore, and Bondy and Chvátal. While Dirac's theorem requires every vertex of a graph G of order $n \geq 3$ to have degree at least $n/2$ in order to guarantee that G is Hamiltonian and Ore's theorem requires many vertices to have degree at least $n/2$, neither theorem can be applied if the maximum degree of G is less than $n/2$.

There are some sufficient conditions for an r-regular graph G to be Hamiltonian that do not require $r \geq |V(G)|/2$. One of these is due to Crispin Nash-Williams [52].

Fig. 2.4 The Petersen graph

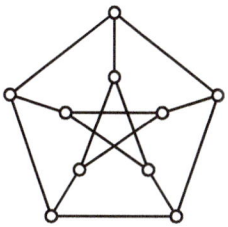

Theorem 2.5 (Nash-Williams' Theorem). *Every r-regular graph of order $2r + 1 \geq 5$ is Hamiltonian.*

Proof. Let G be an r-regular graph of order $2r + 1$, where then r is a positive even integer. Since C_5 is the unique 2-regular graph of order 5, suppose that $r \geq 4$. By Theorem 2.3 (a), we may assume that G contains a Hamiltonian path, say $P = (v_0, v_1, \ldots, v_{2r})$. Suppose, however, that G is not Hamiltonian. For $1 \leq i \leq 2r$, it follows that if $v_0 v_i \in E(G)$, then $v_{i-1} v_{2r} \notin E(G)$; for otherwise, $(v_0, v_i, v_{i+1}, \ldots, v_{2r}, v_{i-1}, v_{i-2}, \ldots, v_0)$ is a Hamiltonian cycle of G, which is impossible. Since $\deg_G v_0 = \deg_G v_{2r} = r$, it follows that exactly one of $v_0 v_i$ and $v_{i-1} v_{2r}$ belongs to $E(G)$ for $1 \leq i \leq 2r$. We consider the following two cases.

Case 1. $N(v_0) = \{v_1, v_2, \ldots, v_r\}$. Then $N(v_{2r}) = \{v_r, v_{r+1}, \ldots, v_{2r}\}$. If $G - v_r$ is disconnected, then $G - v_r$ is a subgraph of $2K_r$. However then, either $\delta(G) < r$ or $\deg_G v_r = 2r$, which contradicts the fact that G is r-regular. Thus, v_r is not a cut-vertex and so $v_{i*} v_{j*} \in E(G)$ for some i^* and j^* satisfying $1 \leq i^* \leq r - 1$ and $r + 1 \leq j^* \leq 2r - 1$. However, then, this produces a Hamiltonian cycle $(v_0, v_1, \ldots, v_{i*}, v_{j*}, v_{j*+1}, \ldots, v_{2r}, v_{j*-1}, v_{j*-2}, \ldots, v_{i*+1}, v_0)$. Thus, this case never occurs.

Case 2. $N(v_0) \neq \{v_1, v_2, \ldots, v_r\}$. Then there exists an integer i^* such that $v_0 v_{i*} \notin E(G)$ and $v_0 v_{i*+1} \in E(G)$. Thus, $v_{2r} v_{i*-1} \in E(G)$ and $v_{2r} v_{i*} \notin E(G)$ and so we have a $2r$-cycle $C = (v_0, v_1, \ldots, v_{i*-1}, v_{2r}, v_{2r-1}, \ldots, v_{i*+1}, v_0)$. Renaming the vertices, let us write $C = (u_1, u_2, \ldots, u_{2r}, u_1)$. Since $\deg v_{i*} = r$ and G is not Hamiltonian, it follows that either $N(v_{i*}) = \{u_1, u_3, \ldots, u_{2r-1}\}$ or $N(v_{i*}) = \{u_2, u_4, \ldots, u_{2r}\}$, say the former. Then for each u_{2i} ($1 \leq i \leq r$), the cycle obtained from C by replacing u_{2i} by v_{i*} is also a $2r$-cycle. This implies that $N(v_{i*}) = N(u_{2i})$ for $1 \leq i \leq r$. It then follows that $\deg u_1 \geq r + 1$, which is again a contradiction. $\square$

If G is r-regular and $|V(G)| \geq 2r + 2$, then G may or may not be connected. (Consider, for example, $G = 2K_{r+1}$.) Jackson [43] has shown that if G is 2-connected and its order is at most $3r$, then G is guaranteed to be Hamiltonian. In fact, Zhu, Liu, and Yu [72] showed that $3r$ can be replaced by $3r + 1$ by excluding the *Petersen graph* (see Fig. 2.4).

Theorem 2.6. *Every 2-connected r-regular graph of order at most* $3r + 1$ *is Hamiltonian unless it is the Petersen graph.*

The following is therefore a consequence of the above theorem.

Theorem 2.7. *Let r be a positive integer. If G is an r-regular graph of order* $2r + 2$, *then either G is Hamiltonian or* $G = 2K_{r+1}$.

Proof. Let G be an r-regular graph of order $2r + 2$. If $r = 1$, then clearly $G = 2K_2$. For $r \geq 2$, it suffices to show by Theorem 2.6 that if G is not 2-connected, then $G = 2K_{r+1}$. Since G is disconnected if and only if $G = 2K_{r+1}$, we may assume that G is connected and has a cut-vertex v. Clearly G is not complete. Also, G being r-regular implies that each component of $G - v$ contains at least r vertices. Thus, $G - v$ consists of exactly two components, say G_1 and G_2, whose orders are r and $r + 1$, respectively. However then, the vertices in G_1 have degree r in G only if $N(v) = V(G_1)$ and $G_1 = K_r$, which then implies that G is already disconnected even without deleting v. $\square$

Another sufficient condition for a graph to be Hamiltonian is due to Chvátal and Paul Erdős [27]. Before presenting their result, let us state a few lemmas.

For a graph G that is not a forest, the length of a longest cycle in G is called the *circumference* cir(G) of G. Therefore, G is Hamiltonian if and only if cir(G) = $|V(G)| \geq 3$.

Lemma 2.1. *If G is a graph with* $\delta(G) \geq 2$, *then G contains cycles and* cir(G) $\geq$ $\delta(G) + 1$.

Proof. That G is not a forest is immediate. Suppose that $P = (v_0, v_1, v_2, \ldots, v_\ell)$ is a longest path in G. Since P cannot be extended any further, the neighborhood of v_0 must be a subset of $V(P) \backslash \{v_0\}$. Thus, there exists an integer ℓ' satisfying $\delta(G) \leq \deg v_0 \leq \ell' \leq \ell$ such that $v_{\ell'}$ belongs to P and $v_0 v_{\ell'} \in E(G)$. Thus, $(v_0, v_1, \ldots, v_{\ell'-1}, v_{\ell'}, v_0)$ is a cycle in G whose length is at least $\delta(G) + 1$. $\square$

For a graph G, consider a subset $S \subseteq V(G)$. The set S is *independent* if no two vertices in S are adjacent in G. The *independence number* $\alpha(G)$ of G is the maximum number of vertices in an independent set of vertices of G. If G is not complete and $G - S$ is disconnected, then the set S is called a *vertex-cut* of G. The cardinality of a minimum vertex-cut of G is the *connectivity* of G, denoted by $\kappa(G)$. When G is a complete graph, its connectivity is defined as $|V(G)| - 1$. Observe that $1 \leq \alpha(G) \leq |V(G)|$ while $0 \leq \kappa(G) \leq |V(G)| - 1$. If k is a positive integer satisfying $k \leq \kappa(G)$, then G is said to be k-*connected*. In other words, for a k-connected graph G, deleting $k - 1$ arbitrary vertices from G does not disconnect the graph. See [24, p. 92], for example, for the proof of the following well-known result.

Theorem 2.8. *For every graph G, $\kappa(G) \leq \delta(G)$.*

Theorem 2.9 (Chvátal-Erdős' Theorem). *Let G be a k-connected graph of order at least 3. If $k \geq \alpha(G)$, then G is Hamiltonian.*

Proof. Suppose that G is a k-connected graph containing more than two vertices, where $k \geq \alpha(G)$, and assume, to the contrary, that G is not Hamiltonian. Since $\alpha(G) = 1$ if and only if G is a complete graph, which is Hamiltonian, we may assume that $\alpha(G) \geq 2$. Lemma 2.1 and Theorem 2.8 then imply that $3 \leq k + 1 \leq$ cir$(G) \leq |V(G)| - 1$. Suppose that $C = (v_1, v_2, \ldots, v_\ell, v_{\ell+1} = v_1)$ is a cycle in G whose length is $\ell = $ cir(G) and let H be a component in $G - V(C)$.

Consider the subsets S and S' of $V(C)$ such that, for $1 \leq i \leq \ell$, $v_i \in S$ if and only if v_i is adjacent to a vertex in H if and only if $v_{i+1} \in S'$. Then S is nonempty since G is connected. Note also that if S contains two distinct vertices, say u and v, then G contains a $u - v$ path of length at least 2 each of whose internal vertices is a vertex in H. Therefore, by the fact that C is a longest cycle in G, no two consecutive vertices on C belong to S, that is, $S \cap S' = \emptyset$. Thus, S is a vertex-cut of G and so $|S'| = |S| \geq k$.

We now verify that S' is an independent set. If this is not the case, then there are integers i and j satisfying $1 \leq i < j \leq \ell$ such that $v_{i+1}, v_{j+1} \in S'$ and $v_{i+1}v_{j+1} \in E(G)$. Then G contains the $v_i - v_j$ path P, where

$$P = \begin{cases} (v_1, v_\ell, v_{\ell-1}, \ldots, v_{j+1}, v_2, v_3, \ldots, v_j) & \text{if } i = 1 \\ (v_i, v_{i-1}, \ldots, v_1, v_{i+1}, v_{i+2}, \ldots, v_\ell) & \text{if } j = \ell \\ (v_i, v_{i-1}, \ldots, v_1, v_\ell, v_{\ell-1}, \ldots, v_{j+1}, v_{i+1}, v_{i+2}, \ldots, v_j) & \text{otherwise.} \end{cases}$$

In each case, $V(P) = V(C)$. On the other hand, since both v_i and v_j belong to S, there is also a $v_i - v_j$ path Q of length at least 2 in G such that $V(C) \cap V(Q) = \{v_i, v_j\}$. However, this is impossible since P and Q form a cycle in G whose length is at least $\ell + 1 = $ cir$(G) + 1$. Thus, S' is independent, as claimed. Furthermore, for an arbitrary vertex $x \in V(H)$, the set $S' \cup \{x\}$ is independent. However then, $k + 1 \leq |S' \cup \{x\}| \leq \alpha(G)$, which contradicts our original assumption that $k \geq \alpha(G)$. $\square$

Therefore, a graph of order at least 3 must be Hamiltonian provided its connectivity is at least as large as its independence number. The complete bipartite graph $K_{n,n+1}$ ($n \geq 1$) shows that the bound is sharp as $\kappa(K_{n,n+1}) = n = \alpha(K_{n,n+1}) - 1$. Note that $K_{n,n+1}$ is traceable although it is not Hamiltonian.

If G is a nontrivial k-connected graph with $k \geq \alpha(G) - 1$, then consider the graph $H = G \vee K_1$, the join of G and a vertex. One can verify that H is $(k + 1)$-connected and $\alpha(H) = \alpha(G)$. Hence, Theorem 2.9 guarantees that H contains a Hamiltonian cycle, which in turn implies that G contains a Hamiltonian path.

Theorem 2.10 ([27]). *If G is a nontrivial k-connected graph, where $k \geq \alpha(G) - 1$, then G is traceable.*

2.3   The Toughness of a Graph

While a number of sufficient conditions have been derived for a graph to be Hamiltonian, there is one well-known and useful necessary condition. We have already stated that every Hamiltonian graph is 2-connected, that is, no Hamiltonian graph contains a cut-vertex. Stated in another manner, no Hamiltonian graph G contains a vertex v such that $G - v$ contains two or more components. In fact, every Hamiltonian graph satisfies an even more general condition. The number of components in a graph G is denoted by $k(G)$.

Theorem 2.11. *Let G be a Hamiltonian graph. Then $k(G - S) \le |S|$ for every nonempty proper subset S of $V(G)$.*

Proof. Let S be a nonempty proper subset of $V(G)$. Suppose that $k(G - S) = k \ge 2$ and that $G_1, G_2, \ldots, G_k$ are the k components of $G - S$. Therefore, each vertex in G_i can only be adjacent to vertices in S or to other vertices in G_i. Let $C = (v_1, v_2, \ldots, v_n, v_{n+1} = v_1)$ be a Hamiltonian cycle in G, where $n = |V(G)|$, and let $i_j = \max\{i : v_i \in V(G_j), 1 \le i \le n\}$ for $1 \le j \le k$. Thus, the set $\{v_{i_1+1}, v_{i_2+1}, \ldots, v_{i_k+1}\}$ is a subset of S containing k distinct vertices. It then follows that $|S| \ge k = k(G - S)$.   □

Necessary conditions are typically most useful when stated in their contrapositive forms.

Theorem 2.12. *If G is a graph containing a nonempty proper subset S of $V(G)$ such that $k(G - S) > |S|$, then G is not Hamiltonian.*

As a consequence of Theorem 2.11, if G is Hamiltonian, then $|S|/k(G - S) \ge 1$ for every nonempty proper subset S of $V(G)$. This observation led Chvátal to introduce a new concept in 1973.

For a positive real number t, a noncomplete graph G is t-*tough* if

$$\frac{|S|}{k(G - S)} \ge t$$

for every vertex-cut S of G. The *toughness* $t(G)$ of G is the maximum real number t for which G is t-tough. For a complete graph K_n, its toughness is taken as $t(K_n) = (n - 1)/2$.

By our earlier observations, every Hamiltonian graph is 1-tough. The converse is not true, however. For example, the graph G of Fig. 2.5 is 1-tough but is not Hamiltonian. In addition, it is well known that the Petersen graph P is not Hamiltonian; yet P is not only 1-tough, it is $(4/3)$-tough. In 1973, Chvátal [26] made the following conjecture.

Fig. 2.5 A non-Hamiltonian
1-tough graph

Fig. 2.6 Constructing the
Bauer-Broersma-Veldman
graph

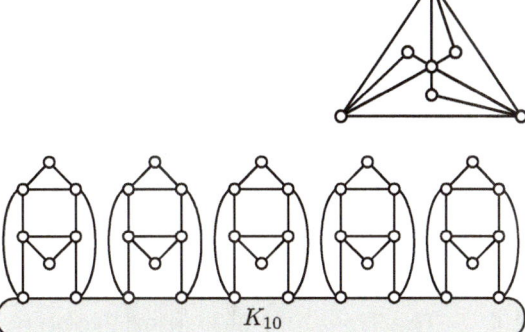

K_{10}

Chvátal's Toughness Conjecture. *There exists a real number t_0 such that every t_0-tough graph is Hamiltonian.*

In 1985, Enomoto, Jackson, Katernis, and Saito [31] proved that every 2-tough graph contains a 2-factor (a spanning 2-regular subgraph) and that there is no real number $t < 2$ for which every t-tough graph contains a 2-factor. This added credence to the following.

The 2-Tough Conjecture. *Every 2-tough graph is Hamiltonian.*

However, in 2000, Bauer, Broersma, and Veldman [9] showed that the 2-Tough Conjecture is false by constructing the so-called *Bauer-Broersma-Veldman graph*, a graph that is 2-tough but not Hamiltonian. This graph is formed by taking the join of K_2 and the graph of order 40 shown in Fig. 2.6. In fact, Bauer, Broersma, and Veldman established the following.

Theorem 2.13 ([9]). *For every real number $t < 9/4$, there is a t-tough nontraceable graph.*

Hamiltonian-Connected Graphs

A graph G is *Hamiltonian-connected* if G contains a Hamiltonian $u - v$ path for every two distinct vertices u and v of G. Thus, every Hamiltonian-connected graph of order at least 3 is Hamiltonian (but the converse is of course false as cycles show). In fact, every edge in a Hamiltonian-connected graph belongs to a Hamiltonian cycle in that graph.

A number of conditions for a graph to be Hamiltonian-connected are known that are similar to those for a graph to be Hamiltonian. The following results are analogous to Theorems 2.1, 2.2, 2.12, and 2.9, respectively.

Theorem 2.14 ([58]). *If G is a graph of order $n \geq 4$ such that* $\deg u + \deg v \geq n+1$ *for every pair u, v of nonadjacent vertices of G, then G is Hamiltonian-connected. Consequently, $\delta(G) \geq (n+1)/2$ implies that G is Hamiltonian-connected.*

Theorem 2.15. *If G is a graph containing a nonempty proper subset S of $V(G)$ such that $k(G - S) > |S| - 1$, then G is not Hamiltonian-connected.*

Theorem 2.16 ([27]). *If G is a k-connected graph satisfying $k \geq \alpha(G) + 1$, then G is Hamiltonian-connected.*

2.4   The Traveling Salesman Problem

One of the best known problems concerning Hamiltonian cycles is of a more applied nature.

The Traveling Salesman Problem. *A salesman wishes to make a round trip that visits certain cities once each. He knows the distance between each pair of cities. What is the minimum total distance of such a round trip?*

This problem can be expressed in terms of weighted graphs. In particular, let G be a weighted complete graph whose vertices are the cities and where each edge uv is assigned a weight equal to the distance between u and v. The *weight* $w(C)$ of a Hamiltonian cycle C in G is the sum of the weights of the edges of C. Finding a solution to the Traveling Salesman Problem then consists of determining the minimum weight of a Hamiltonian cycle in G.

If the number n of cities involved is large, then the number of Hamiltonian cycles in G that need to be investigated is quite large. We can consider a Hamiltonian cycle as beginning at any vertex v. Then the remaining $n - 1$ vertices can follow v in any of $(n - 1)!$ orders. This produces $(n - 1)!$ Hamiltonian cycles whose weights we need to compute. In fact, we need *only* consider $(n - 1)!/2$ Hamiltonian cycles as we would obtain the same sum if the order in which the vertices appear in a cycle were reversed.

Even though the Traveling Salesman Problem is an extremely difficult problem in general, there are instances where this problem has been solved for a large number of cities. In 1998 Applegate, Bixby, Chvátal, and Cook [5] solved a Traveling Salesman Problem for the 13,509 largest cities in the United States (those whose population exceeded 500 at that time). They also solved a Traveling Salesman Problem for 15,113 German cities in 2001 and for 24,978 Swedish cities in 2004. Their ultimate goal was to solve the Traveling Salesman Problem for every registered city or town in the world plus a few research bases in Antartica (1,904,711 locations in all). In 2006, the four wrote a book titled *The Traveling Salesman Problem: A Computational Study* [6], in which they describe the history of the Traveling Salesman Problem as well as the method they used to solve a range of large-scale problems. In 2012 Cook [28] wrote a book titled *In Pursuit of the Traveling Salesman* for a more general audience.

2.5   Line Graphs and Powers of Graphs

There are two operations on graphs in which much attention has been focused regarding Hamiltonian properties of the resulting graphs.

Line Graphs

The *line graph* $L(G)$ of a nonempty graph G is that graph whose vertices correspond to the edges of G where two vertices of $L(G)$ are adjacent if and only if the corresponding edges of G are adjacent. The line graph of a graph G is therefore Hamiltonian provided the m edges of G can be listed as $e_1, e_2, \ldots, e_m, e_{m+1} = e_1$ in such a way that e_i and e_{i+1} are adjacent for $i = 1, 2, \ldots, m$. As a consequence of this observation, we have the following.

Theorem 2.17. *The line graph of an Eulerian graph is Hamiltonian.*

Each of the three graphs G_1, G_2, G_3 in Fig. 2.7 is neither Eulerian nor Hamiltonian, while $L(G_1)$ is Hamiltonian but not Eulerian, $L(G_2)$ is Eulerian but not Hamiltonian, and $L(G_3)$ is neither Eulerian nor Hamitonian. The graph G_1 shows that the converse of Theorem 2.17 is not true. In fact, Harary and Nash-Williams [39] characterized those graphs whose line graphs are Hamiltonian. A circuit C in a graph G is called a *dominating circuit* if every edge of G is incident with at least one vertex of C.

Theorem 2.18. *Let G be a graph without isolated vertices. Then $L(G)$ is Hamiltonian if and only if either G is a star of size at least 3 or G contains a dominating circuit.*

Proof. Let m be the size of G. If G is the star $K_{1,m}$, then $L(G) = K_m$, which is Hamiltonian for $m \geq 3$. Suppose then that G contains a dominating circuit $C = (v_1, v_2, \ldots, v_\ell, v_1)$. It suffices to show that there exists an ordering $s : e_1, e_2, \ldots, e_m$ of the m edges of G such that e_i and e_{i+1} are adjacent edges of G, for $1 \leq i \leq m-1$, as are e_1 and e_m, since such an ordering s corresponds to a Hamiltonian cycle of $L(G)$. Begin s by selecting, in any order, all edges of G incident with v_1 that are not edges of C, followed by the edge v_1v_2. At each successive vertex v_i, $2 \leq i \leq \ell - 1$, select, in any order, all edges of G incident with v_i that are neither edges of C nor previously selected edges, followed by the edge v_iv_{i+1}. This process terminates with the edge $v_{\ell-1}v_\ell$. The ordering s is completed by adding the edge $v_\ell v_1$. Since C is a dominating circuit of G, every edge of G appears exactly once in s. Furthermore, consecutive edges of s as well as the first and last edges of s are adjacent in G.

Fig. 2.7 Graphs and their
line graphs

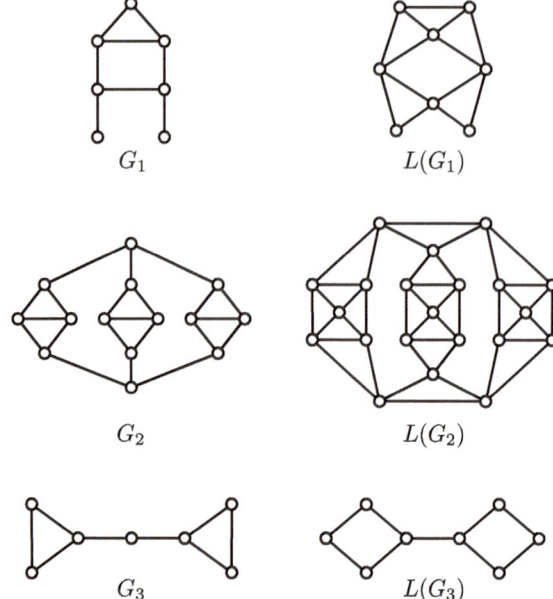

G_1 $L(G_1)$

G_2 $L(G_2)$

G_3 $L(G_3)$

Conversely, suppose that G is not a star but $L(G)$ is Hamiltonian. We show that G contains a dominating circuit. Since $L(G)$ is Hamiltonian, there is an ordering $s : e_1, e_2, \ldots, e_m$ of the m edges of G such that e_i and e_{i+1} are adjacent edges of G for $1 \le i \le m-1$, as are e_1 and e_m. For $1 \le i \le m-1$, let v_i be the vertex of G incident with both e_i and e_{i+1}. (Note that $1 \le i \ne j \le m-1$ does not necessarily imply that $v_i \ne v_j$.) Since G is not a star, there is a smallest integer i_1 exceeding 1 such that $v_{i_1} \ne v_1$. Thus, the edges $e_1, e_2, \ldots, e_{i_1-1}$ are incident with v_1 and $e_{i_1} = v_1 v_{i_1}$. Next, let i_2 (if it exists) be the smallest integer exceeding i_1 such that $v_{i_2} \ne v_{i_1}$. Then the edges $e_{i_1}, e_{i_1+1}, \ldots, e_{i_2-1}$ are incident with v_{i_1} and $e_{i_2} = v_{i_1} v_{i_2}$. Continuing in this fashion, we finally arrive at a vertex v_{i_ℓ} such that $e_{i_\ell} = v_{i_{\ell-1}} v_{i_\ell}$, where $v_{i_\ell} = v_{m-1}$. Note that (i) v_1 is incident with the edges $e_1, e_2, \ldots, e_{i_1}$, (ii) v_{i_j} $(1 \le j \le \ell-1)$ is incident with the edges $e_{i_j}, e_{i_j+1}, \ldots, e_{i_{j+1}}$, and (iii) v_{i_ℓ} is incident with the edges $e_{i_\ell}, e_{i_\ell+1}, \ldots, e_m$. Since each edge of G appears exactly once in s and $1 < i_1 < i_2 < \cdots < i_\ell \le m-1$, we obtain a trail

$$T = (v_1, v_{i_1}, v_{i_2}, \ldots, v_{i_\ell} = v_{m-1}) = (e_{i_1}, e_{i_2}, \ldots, e_{i_\ell})$$

in G with the properties that every edge of G is incident with a vertex in T and neither e_1 nor e_m belongs to T. Thus, T itself is a dominating circuit if $v_1 = v_{m-1}$. If not, let v_0 be the vertex of G incident with both e_1 and e_m. Now, if $v_0 \notin \{v_1, v_{m-1}\}$, then $(e_{i_1}, e_{i_2}, \ldots, e_{i_\ell}, e_m, e_1)$ is a dominating circuit. Otherwise, $v_0 = v_1$ or $v_0 = v_{m-1}$. If $v_0 = v_1$, then $e_m = v_1 v_{m-1}$ and so $(e_{i_1}, e_{i_2}, \ldots, e_{i_\ell}, e_m)$ is a dominating circuit. Similarly, if $v_0 = v_{m-1}$, then $(e_{i_1}, e_{i_2}, \ldots, e_{i_\ell}, e_1)$ is a dominating circuit.   □

It is an immediate corollary of Theorem 2.18 that $L(G)$ is Hamiltonian whenever G is a graph that is either Eulerian or Hamiltonian. In fact, $L(G)$ is Hamiltonian if G contains an Eulerian spanning subgraph. Another consequence of Theorem 2.18 is that if $L(G)$ is Hamiltonian, then every bridge in G must be a pendant edge. Indeed, if G contains a bridge $e = uv$ where neither u nor v is an end-vertex, then the vertex in $L(G)$ corresponding to e is a cut-vertex. (See the graphs G_1 and G_3 in Fig. 2.7, for example.)

For nearly every connected graph, successively taking the line graphs results in Hamiltonian graphs. For a nonempty graph G, we write $L^0(G)$ for G and $L^1(G)$ for $L(G)$. By $L^2(G)$, we mean $L(L(G))$. More generally, for a positive integer k, the graph $L^k(G)$ is defined as $L(L^{k-1}(G))$.

If G is connected and r-regular, then $L(G)$ is a connected $2(r-1)$-regular graph, that is, $L(G)$ is Eulerian. In this case, $L(G)$ is Hamiltonian if r is even but the same cannot be guaranteed when r is odd. (Consider, for example, the graph G_2 in Fig. 2.7 and the complete graph K_4. Both are 3-regular and $L(G_2)$ is not Hamiltonian while $L(K_4) = K_{2,2,2}$ is.) Taking the line graph again, however, $L^2(G)$ is a $2(2r-3)$-regular graph that is both Eulerian and Hamiltonian for every integer $r \geq 2$. That is, $L^2(G)$ is always both Eulerian and Hamiltonian when G is a connected regular graph containing three or more vertices. For those connected graphs that are not regular, we have the following result due to Chartrand and Wall [17].

Theorem 2.19. *If G is a connected graph with $\delta(G) \geq 3$, then $L^2(G)$ is Hamiltonian.*

Proof. Let v be a vertex of G. Then the edges incident with v in G give rise to a subgraph G_v in $L(G)$ which is isomorphic to a complete graph whose order equals $\deg v$ (≥ 3). Let H_v be a Hamiltonian cycle in G_v and define the spanning subgraph H of $L(G)$ by $V(H) = V(L(G))$ and $E(H) = \bigcup_{v \in V(G)} E(H_v)$. Then H is connected and certainly has a cycle decomposition. Thus, H has an Eulerian circuit, which is a dominating circuit of $L(G)$. The desired result now follows by Theorem 2.18.   □

The graphs G_2 and G_3 in Fig. 2.7 show that Theorem 2.19 cannot be improved in general, as neither $L(G_2)$ nor $L^2(G_3)$ is Hamiltonian.

Note that $L(P_n) = P_{n-1}$ for each integer $n \geq 2$. Thus $L^{n-1}(P_n)$ is trivial and $L^k(P_n)$ is not defined for $k \geq n$. Also, $L(C_n) = C_n$ for every $n \geq 3$ and $L(K_{1,3}) = C_3$. Therefore, $L^k(C_n) = C_n$ for $k \geq 0$; while $L^k(K_{1,3}) = C_3$ for $k \geq 1$. If, however, G is a connected graph that is none of a path, cycle, and the star $K_{1,3}$ (called a *claw*), then we eventually arrive at some positive integer k such that $\deg v \geq 3$ for every vertex v of $L^k(G)$. The following is due to Chartrand [15].

Theorem 2.20. *If G is a connected graph that is not a path, then there exists a positive integer K such that $L^k(G)$ is Hamiltonian for every integer $k \geq K$.*

Powers of Graphs

Another operation on graphs is the k-th power of a graph for various positive integers k, a topic discussed in Sect. 1.2. Recall for a connected graph G and a positive integer k that the k-th power G^k of G is the graph with $V(G^k) = V(G)$ and $E(G^k) = \{uv : 1 \le d_G(u, v) \le k\}$.

If G is connected, then G^k is complete if and only if $k \ge \text{diam}(G)$. Thus, it suffices to consider G^k only when $1 \le k < \text{diam}(G)$. Since G is a spanning subgraph of G^k for every positive integer k, the graph G^k is certainly Hamiltonian if G itself is. For a connected graph G of order $n \ge 3$, there is a smallest positive integer k such that G^k is Hamiltonian. That G^3 is Hamiltonian for every connected graph of order 3 or more is a consequence of a result of Sekanina [61]. Recall that a graph G is *Hamiltonian-connected* if G contains a Hamiltonian $u - v$ path for every two distinct vertices u and v of G.

Theorem 2.21. *The cube of every connected graph is Hamiltonian-connected.*

Proof. If H is a spanning subgraph of G and H^3 is Hamiltonian-connected, then G^3 is also Hamiltonian-connected. Hence, it suffices to prove that the cube of every tree is Hamiltonian-connected. We proceed by induction on n, the order of the tree. Since the result is obvious for those graphs having diameter at most 3, assume for every tree of order less than n that its cube is Hamiltonian-connected for some $n \ge 5$. Let T be a tree of order n. For two arbitrary distinct vertices $u, v \in V(T)$, let $(u = v_1, v_2, \ldots, v_{d+1} = v)$ be the unique $u - v$ path in T, where $d = d_T(u, v)$. Also, let T_1 and T_2 be the two components of $T - v_1 v_2$, where v_i belongs to T_i for $i = 1, 2$. Thus, for each tree T_i, either T_i is trivial or T_i^3 is Hamiltonian-connected. We consider the following two cases.

Case 1. $v = v_2$, that is, $uv \in E(T)$. For each $i = 1, 2$, let $w_i \in N_{T_i}(v_i)$ if T_i is nontrivial and let $w_i = v_i$ otherwise. (Note that at most one of T_1 and T_2 is trivial.) Then $d_T(w_1, w_2) \le 3$ and so $w_1 w_2 \in E(T^3)$. If we let $P^{(i)}$ be a Hamiltonian $v_i - w_i$ path in T_i^3 (which may be trivial) for $i = 1, 2$, then $P^{(1)}$ and $P^{(2)}$ with the edge $w_1 w_2$ form a Hamiltonian $u - v$ path in T^3.

Case 2. $v \ne v_2$. Then T_2^3 contains a Hamiltonian $v_2 - v$ path P. Let $P^{(1)}$ be a Hamiltonian $v_1 - w_1$ path in T_1^3, as described in Case 1. Since $d_T(w_1, v_2) \le 2$, the paths $P^{(1)}$ and P with the edge $w_1 v_2$ form a Hamiltonian $u - v$ path in T^3. □

The following is therefore immediate by the previous result.

Theorem 2.22. *The cube of every connected graph of order at least 3 is Hamiltonian.*

The graph of order 7 in Fig. 2.5 is the square of the tree obtained by subdividing each edge of $K_{1,3}$ exactly once. We have already seen that this graph is not Hamiltonian. Consequently, even though the cube of every connected graph of order at least 3 is Hamiltonian, such is not the case for the square. On the other hand, in the 1960s, Plummer and Nash-Williams independently conjectured that the square of every 2-connected graph is Hamiltonian. In 1974, this conjecture was verified by Fleischner [33].

Theorem 2.23. *The square of every 2-connected graph is Hamiltonian.*

2.6   Hamiltonian Walks and Cyclic Orderings

Let G be a nontrivial connected graph. By a *Hamiltonian walk* in G is meant a closed spanning walk of minimum length in G. Thus, while an Eulerian walk is a closed *edge-covering* walk, not necessarily of minimum length, a Hamiltonian walk is a closed *vertex-covering* walk of *minimum* length. The length of a Hamiltonian walk in G is called the *Hamiltonian number* of G and is denoted by $h(G)$. Therefore, $h(G) \geq |V(G)|$ and $h(G) = |V(G)|$ if and only if G is either Hamiltonian or K_2.

For a connected graph G, recall that $e(G)$ denotes the minimum length of an Eulerian walk in G. We saw in Sect. 1.4 that $|E(G)| \leq e(G) \leq 2|E(G)|$. Therefore, if G is a nontrivial connected graph, then

$$|V(G)| \leq h(G) \leq e(G) \leq 2|E(G)|. \qquad (2.1)$$

That the upper bound $2|E(G)|$ for $h(G)$ cannot be improved is shown in the next result due to Goodman and Hedetniemi [37].

Theorem 2.24. *If T is a tree of order $n \geq 2$, then $h(T) = 2(n-1)$.*

Proof. Since the size of a tree of order n is $n - 1$, it suffices to show by (2.1) that $h(T) \geq 2(n - 1)$. Let W be a Hamiltonian walk in T and consider an edge $uv \in E(T)$. We may assume that u precedes v on W. Since uv is a bridge, it lies on W. We may therefore assume that W begins with u and is immediately followed by v. Since W terminates at u, the vertex u appears a second time on W and this occurrence of u is immediately preceded by v. Thus the edge uv appears at least twice on W. Hence $h(T) \geq 2(n - 1)$ and therefore $h(T) = 2(n - 1)$.   □

The proof of Theorem 2.24 in fact shows that every bridge in a connected graph G must appear at least twice on any Hamiltonian walk in G. Since a Hamiltonian walk in a spanning tree T of G is also a Hamiltonian walk in G, it follows that $h(G) \leq h(T)$. Thus we have the following.

Fig. 2.8 Illustrating cyclic
orderings of the vertices in a
graph

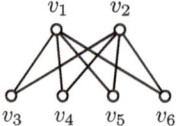

Theorem 2.25. *If G is a nontrivial connected graph of order n, then*

$$n \leq h(G) \leq 2(n-1).$$

In [23] Chartrand, Thomas, Saenpholphat, and Zhang described an alternative way to compute the Hamiltonian number of a graph. If a graph G of order n is Hamiltonian, then each Hamiltonian walk in G is a Hamiltonian cycle C in G, say $C = (v_1, v_2, \ldots, v_n, v_1)$, and so $h(G) = n$. Since the edges $v_1 v_2, v_2 v_3, \ldots, v_{n-1} v_n, v_n v_1$ belong to G, it follows that there is a *cyclic ordering* $v_1, v_2, \ldots, v_n, v_{n+1} = v_1$ of the vertices of G such that $\sum_{i=1}^{n} d(v_i, v_{i+1}) = n$.

In general, for a connected graph G of order $n \geq 2$ and a cyclic ordering $s : v_1, v_2, \ldots, v_n, v_{n+1} = v_1$ of the vertices of G, the number $d(s)$ is defined as

$$d(s) = \sum_{i=1}^{n} d(v_i, v_{i+1}). \tag{2.2}$$

Since $d(v_i, v_{i+1}) \geq 1$ for $i = 1, 2, \ldots, n$, it follows that $d(s) \geq n$. To illustrate this concept, consider the graph $G = K_{2,4}$ shown in Fig. 2.8. The distance between every two vertices of G is either 1 or 2. In every cyclic ordering of the vertices of G, there are either two pairs or four pairs of consecutive vertices with distance 2. Consider, for example, the two cyclic orderings $s_1 : v_1, v_3, v_2, v_4, v_5, v_6, v_1$ and $s_2 : v_1, v_2, v_3, v_4, v_5, v_6, v_1$. Then $d(s_1) = 8$ and $d(s_2) = 10$.

We define the number $h^*(G) = \min \{d(s)\}$, where the minimum is taken over all cyclic orderings s of the vertices of G. Then $h^*(G) \geq n$ for each connected graph G of order $n \geq 3$ and $h^*(G) = n$ if and only if G is Hamiltonian. In the graph $G = K_{2,4}$ of Fig. 2.8, for every cyclic ordering s of $V(G)$, either $d(s) = 8$ or $d(s) = 10$. Thus $h^*(G) = 8$.

The interest in the parameter $h^*(G)$ lies in the following theorem.

Theorem 2.26. *For every connected graph G, $h^*(G) = h(G)$.*

Proof. The result clearly holds if the order of G is 2 and so we assume that $|V(G)| = n \geq 3$. For a cyclic ordering $s : v_1, v_2, \ldots, v_n, v_{n+1} = v_1$ of $V(G)$ with $d(s) = h^*(G)$, let $P^{(i)}$ be a $v_i - v_{i+1}$ geodesic in G for $1 \leq i \leq n$. Then the walk obtained by proceeding along the paths $P^{(1)}, P^{(2)}, \ldots, P^{(n)}$ in the given order is a closed spanning walk of G whose length equals $h^*(G)$. Therefore, $h(G) \leq h^*(G)$.

Next, let $W = (x_0, x_1, \ldots, x_\ell)$ be a Hamiltonian walk in G. Hence, $L(W) = h(G) \geq n$. Let $v_i = x_{i-1}$ for $i = 1, 2$. For $3 \leq i \leq n$, let $v_i = x_{j_i}$, where j_i is the smallest positive integer such that $x_{j_i} \notin \{v_1, v_2, \ldots, v_{i-1}\}$. Then $s : v_1, v_2, \ldots, v_n, v_{n+1} = v_1$ is a cyclic ordering of $V(G)$. For $1 \le i \le n$, let W_i be the $v_i - v_{i+1}$ subwalk of W. Then

$$h^*(G) \le d(s) = \sum_{i=1}^{n} d(v_i, v_{i+1}) \le \sum_{i=1}^{n} L(W_i) = L(W) = h(G),$$

which completes the proof.   □

By Theorems 2.25 and 2.26, $h^*(T) = h(T) = 2(n-1)$ for every tree T of order n. In fact, trees are the only graphs G satisfying $h(G) = 2(|V(G)| - 1)$.

Theorem 2.27 ([23]). *Let G be a nontrivial connected graph of order n. Then $h(G) = 2(n-1)$ if and only if G is a tree.*

Proof. By Theorem 2.24, it remains to show that $h(G) < 2(n-1)$ if G contains cycles. We proceed by induction on n. If $n = 3$, then $G = K_3$ and the result is immediate. Assume for an integer $n \ge 4$ that $h(H) < 2(n-2)$ for each connected graph H of order $n-1$ that is not a tree. Let G be a connected graph of order n that is not a tree. Since $h(C_n) = n < 2(n-1)$, we may assume that $G \ne C_n$.

We claim that G contains a vertex u such that $G - u$ is connected but not a tree. If G contains cut-vertices, then there is a vertex u in an end-block of G with the desired property. Thus we may assume that G is 2-connected. Since $G \ne C_n$, it follows that G contains a cycle C whose length is less than n. Thus, there is a vertex u that is not a cut-vertex and $G - u$ still contains C. Since $h(G - u) < 2(n-2)$, there is a cyclic ordering $s_0 : v_1, v_2, \ldots, v_{n-1}, v_1$ of the vertices of $G - u$ such that $d(s_0) = h(G - u) < 2(n-2)$. Suppose that u is adjacent to the vertex v_i and consider the cyclic ordering s of $V(G)$ defined by $s : v_1, v_2, \ldots, v_i, u, v_{i+1}, \ldots, v_{n-1}, v_1$. Since $d(u, v_i) = 1$, it follows that $d(u, v_{i+1}) \le d(u, v_i) + d(v_i, v_{i+1}) = 1 + d(v_i, v_{i+1})$. Hence

$$
\begin{aligned}
d(s) &= d(s_0) - d(v_i, v_{i+1}) + d(v_i, u) + d(u, v_{i+1}) \\
&\le d(s_0) - d(v_i, v_{i+1}) + 1 + (1 + d(v_i, v_{i+1})) \\
&< 2(n-2) + 2 = 2(n-1).
\end{aligned}
$$

Therefore, $h(G) \le d(s) < 2(n-1)$.   □

In Theorem 1.23, we saw for a connected graph G of size $m \ge 1$ that $e(G) = 2m$ if and only if G is a tree. It then follows by Theorem 2.24 that if G is a tree of order n, then $h(G) = e(G) = 2(n-1)$. It was shown in [37] that there are connected graphs G that are not trees and yet $h(G) = e(G)$. In order to describe a class of graphs with this property, it is useful to introduce a new term. A cycle C in a connected graph G is a *cut-cycle* of G if $G - E(C)$ is disconnected.

For an Eulerian walk W of a connected graph G, let M be the multigraph obtained from G by replacing each edge uv of G by i parallel edges, where i equals the number of times the edge uv is encountered on W. In this case, M is said to be *induced* by W. In other words, M is the multigraph induced by $E(W)$. Consequently, M is Eulerian and $V(M) = V(G)$.

Theorem 2.28. *If G is a connected graph such that $h(G) = e(G)$, then every cycle of G is a cut-cycle.*

Proof. If G is a tree, then $h(G) = e(G)$ and the result follows vacuously. Otherwise, we may assume, to the contrary, that $h(G) = e(G)$ and G contains a cycle C that is not a cut-cycle. Therefore, $G - E(C)$ is a connected spanning subgraph of G. Suppose that C is an ℓ-cycle. Let W be an Eulerian walk of G with $L(W) = e(G)$ and let M be the multigraph induced by $E(W)$. Then M is an Eulerian multigraph and $V(M) = V(G)$. Since C is not a cut-cycle, $M - E(C)$ is an Eulerian spanning submultigraph of M. An Eulerian circuit in $M - E(C)$ gives rise to a closed spanning walk in $G - E(C)$ and so in G. Hence $h(G) \le |E(M) - E(C)| = e(G) - \ell < e(G)$, which is a contradiction.   □

Theorem 2.28 was strengthened in 1974 by Goodman and Hedetniemi [37].

Theorem 2.29. *If a connected graph G contains a cycle such that more than half of its edges can be removed without disconnecting G, then $h(G) < e(G)$.*

Proof. Let W be an Eulerian walk of G with $L(W) = e(G)$ and let M be the Eulerian multigraph induced by W. Let C be a cycle of G such that $E(C)$ can be partitioned into E_1 and E_2 with $|E_1| > |E_2|$ and $G - E_1$ is connected. Certainly, C is a cycle in M. Also, $M - E_1$ is a connected spanning submiltigraph of M, since $G - E_1 \subseteq M - E_1$ and $G - E_1$ is connected. For each edge $e = uv$ in E_2, we add an additional edge joining u and v in $M - E_1$. This produces an Eulerian multigraph M' whose vertex set is $V(G)$. An Eulerian circuit in M' gives rise to a closed spanning walk in $G - E_1$ and so in G. Hence $h(G) \le |E(M)| - |E_1| + |E_2| = |E(M)| - (|E_1| - |E_2|) < |E(M)| = e(G)$.   □

The converse of Theorem 2.29 is false. For example, consider $G = K_{2,3}$, where $h(G) = 6$ and $e(G) = 8$. In this case, the removal of more than half of the edges of every cycle results in a subgraph of $K_1 + P_4$, which is disconnected. On the other hand, if G is an Eulerian graph, then the converse is true, as the next result shows [37]. In order to present a proof of this result, we first make some preliminary observations. We saw that if W is an Eulerian walk of minimum length in a graph G, then each edge of G appears at most twice in W. It was shown in [37] that this is also the case for a Hamiltonian walk in a graph.

Theorem 2.30. *Every edge in a connected graph G appears at most twice in a Hamiltonian walk in G.*

Proof. Suppose that there exists some edge uv of a connected graph G that appears at least three times in a Hamiltonian walk W in G. We may assume that W has one of the following two forms

$$W' = (u, v, W_1, u, v, W_2, u, v, W_3) \text{ and } W'' = (u, v, W_1, u, v, W_2, v, u, W_3),$$

where W_1, W_2, W_3 are (possibly empty) subwalks in W. Let $\overleftarrow{W}_i$ denote the reverse of the subwalk W_i. If $W = W'$, then the walk $(u, \overleftarrow{W}_1, v, W_2, u, v, W_3)$ is a closed spanning walk of G which is shorter than W. This contradicts the defining property of W. Similarly, if $W = W''$, then the walk $(u, \overleftarrow{W}_1, v, W_2, v, u, W_3)$ is a closed spanning walk of G which is shorter than W, another contradiction. □

The following theorem deals with Hamiltonian walks in Eulerian graphs.

Theorem 2.31. *Let G be an Eulerian graph. Then $h(G) < e(G)$ if and only if G contains a cycle such that more than half of its edges can be removed without disconnecting G.*

Proof. By Theorem 2.29, we only show that an Eulerian graph G with $h(G) < e(G)$ has a cycle the removal of more than half of whose edges from G does not disconnect G. Let G be an Eulerian graph and consider a Hamiltonian walk W of G. By Theorem 2.30, we have a partition $\{E_0, E_1, E_2\}$ of $E(G)$ such that $e \in E_i$ if and only if e appears i times in W for $0 \leq i \leq 2$. Therefore, E_0 is not an edge-cut of G. Also, $h(G) = L(W) = |E_1| + 2|E_2|$ and $e(G) = |E(G)| = |E_0| + |E_1| + |E_2|$. Thus, $h(G) < e(G)$ implies that $|E_0| > |E_2|$.

For a vertex $v \in V(G)$, let E_v be the set of the edges incident with v in G. Of course, $\deg_G v = |E_v|$ is even since G is Eulerian. Also, since the multigraph M induced by $E(W)$ is Eulerian whose vertex set equals $V(G)$, it follows that $\deg_M v = |E_v \cap E_1| + 2|E_v \cap E_2|$ is also even, which in turn implies that $|E_v \cap (E_0 \cup E_2)|$ is even. Thus, the graph G' induced by $E_0 \cup E_2$ is a nonempty spanning subgraph of G in which every nontrivial component is Eulerian. Thus, G' has a cycle decomposition according to Veblen's Theorem and so G' (and G as well) contains a cycle C such that $|E(C) \cap E_0| > |E(C) \cap E_2|$. Now, $G - (E(C) \cap E_0)$ must be connected since E_0 is not an edge-cut of G. □

One of the best known sufficient conditions for a graph to be Hamiltonian is that due to Ore (Theorem 2.2). This theorem can be stated in terms of the Hamiltonian number of a graph as follows.

Theorem 2.32. *If G is a graph of order $n \geq 3$ such that $\deg u + \deg v \geq n$ whenever $uv \notin E(G)$, then $h(G) = n$.*

Jean-Claude Bermond [10] generalized this result by showing that if G is a connected graph of order n for which the minimum degree sum σ of every two nonadjacent vertices of G satisfies $2 \leq \sigma \leq n$, then $h(G)$ is no more than $2n - \sigma$.

Theorem 2.33 (Bermond's Theorem). *Let G be a connected graph G of order $n \geq 3$. If $\deg u + \deg v \geq \sigma$ for every pair u, v of nonadjacent vertices of G and $2 \leq \sigma \leq n$, then $h(G) \leq 2n - \sigma$.*

Among the results obtained by Goodman and Hedetniemi is the following [37].

Theorem 2.34. *Let G be a connected graph having blocks $B_1, B_2, \ldots, B_k$. Then the union of the edges in a Hamiltonian walk for each of the blocks B_i forms a Hamiltonian walk for G and, conversely, the edges in a Hamiltonian walk of G that belong to B_i form a Hamiltonian walk in B_i.*

Theorem 2.34 implies that the study of Hamiltonian walks can be restricted to 2-connected graphs. For k-connected graphs ($k \geq 2$) of a specified diameter, the following appears in [37]. The *diameter* of a connected graph G is the largest distance between two vertices of G and is denoted by diam(G).

Theorem 2.35. *If G is a k-connected graph of order n having diameter d, then*

$$h(G) \leq 2(n - 1) - 2 \lfloor k/2 \rfloor (d - 1).$$

The *clique number* of a graph G is the maximum order among the complete subgraphs of G. In [59] an upper bound was established for $h(G)$ in terms of the order and clique number of a connected graph G.

Theorem 2.36. *If G is a nontrivial connected graph of order n having clique number ω, then $h(G) \leq 2n - \omega$. Furthermore, for each integer ω with $2 \leq \omega \leq n$, there exists a connected graph G of order n having clique number ω such that $h(G) = 2n - \omega$.*

By Theorem 2.27, trees of order n are the only connected graphs of order n with Hamiltonian number $2(n - 1)$. All connected graphs of order n with Hamiltonian number $2n - 3$ or $2n - 4$ are characterized in [59]. A connected graph with exactly one cycle is called a *unicyclic graph*. Therefore, a unicyclic graph is a graph obtained from a tree by joining two nonadjacent vertices. In other words, G is unicyclic if G itself is a cycle or G contains exactly one block that is a cycle and each of the remaining block is K_2.

Theorem 2.37. *Let G be a connected graph of order $n \geq 3$. Then $h(G) = 2n - 3$ if and only if G is a unicyclic graph whose unique cycle is a triangle.*

Let $\mathscr{G}_1$ be the set of connected graphs G of order $n \geq 5$ with cut-vertices such that G contains exactly two blocks that are K_3 and each of the remaining blocks of G is K_2. Also, let $\mathscr{G}_2$ be the set of connected graphs G of order $n \geq 5$ with cut-vertices such that G contains exactly one block that is one of the graphs in the set

$$\{K_4\} \cup \{K_{2,n'-2}, K_{1,1,n'-2} : 4 \leq n' \leq n-1\}$$

and each of the remaining blocks of G is K_2.

Theorem 2.38. *Let G be a connected graph of order n. Then $h(G) = 2n - 4$ if and only if (a) $n \geq 4$ and $G \in \{K_4, K_{2,n-2}, K_{1,1,n-2}\}$ or (b) $n \geq 5$ and $G \in \mathscr{G}_1 \cup \mathscr{G}_2$.*

We have seen that if T is a nontrivial tree of order n, then $h(T) = 2(n-1)$, which is clearly even. With the aid of the alternative definition of the Hamiltonian number of a graph in terms of $d(s)$ defined in (2.2), we can extend this fact to all connected bipartite graphs.

Theorem 2.39 ([35]). *If G is a nontrivial connected bipartite graph, then $d(s)$ is even for every cyclic ordering s of $V(G)$.*

Proof. For an arbitrary cyclic ordering $s : v_1, v_2, \ldots, v_n, v_{n+1} = v_1$ of $V(G)$, where $n = |V(G)|$, consider the set $\{i_1, i_2, \ldots, i_k\}$ of integers with $1 = i_1 < i_2 < \cdots < i_k = n+2$ (where the subscripts of the vertices are expressed as integers modulo n) such that (i) v_{i_j} and $v_{i_{j+1}}$ belong to different partite sets ($1 \leq j \leq k-2$) and (ii) the set $S_j = \{v_i : i_j \leq i < i_{j+1}\}$ is contained in a partite set ($1 \leq j \leq k-1$). Since $v_1 = v_{n+1}$ belongs to both S_1 and S_{k-1}, it follows that k must be even and the partite sets of G are $S_1 \cup S_3 \cup \cdots \cup S_{k-1}$ and $S_2 \cup S_4 \cup \cdots \cup S_{k-2}$. Therefore, $d(v_i, v_{i+1})$ is odd if and only if $i = i_j - 1$ ($2 \leq j \leq k-1$), that is, exactly $k-2$ of the n summands in $d(s)$ are odd. $\square$

Alternatively, we may consider Theorem 2.39 as follows. For a nontrivial connected graph G of order n, suppose that $s : v_1, v_2, \ldots, v_n, v_{n+1} = v_1$ is a cyclic ordering of $V(G)$. If $P^{(i)}$ is a $v_i - v_{i+1}$ geodesic for $1 \leq i \leq n$, then the walk W obtained by traversing the n paths $P^{(1)}, P^{(2)}, \ldots, P^{(n)}$ in this order is a closed walk in which every vertex of G appears at least once. Furthermore, the length of W equals $d(s)$. Since the length of a $u - v$ walk in a bipartite graph is even if and only if u and v belong to the same partite set, every closed walk in a bipartite graph has even length.

Theorem 2.40. *If G is a connected bipartite graph, then $h(G)$ is even.*

Hamiltonian walks in maximal planar graphs were studied by Asano, Nishizeki, and Watanabe [7, 8]. In [7], it was shown that if G is a maximal planar graph of order $n \leq 10$, then G is Hamiltonian and so $h(G) = n$. For a maximal planar graph of order $n \geq 11$, an upper bound was established in terms of n.

Theorem 2.41. *If G is a maximal planar graph of order $n \geq 11$, then $h(G) \leq 1.5(n - 3)$.*

As indicated in [8], the problem of finding a Hamiltonian walk in a given graph is NP-complete. This problem is a generalized Hamiltonian cycle problem and is a special case of the Traveling Salesman Problem. With the aid of the techniques of divide-and-conquer and augmentation, an approximation algorithm for this problem on maximal planar graphs was presented in [8]. This algorithm finds in $O(n^2)$ time, a closed spanning walk of length at most $3(n - 3)/2$ in a given arbitrary maximal planar graph of order $n \geq 9$. More recent results include the following by Kawarabayashi and Ozeki [44].

Theorem 2.42. *Let G be a 3-connected planar graph. Then $h(G) \leq 4(n - 1)/3$, where $|V(G)| = n$.*

2.7   The Upper Hamiltonian Number of a Graph

In Sect. 2.6, we saw for the graph $G = K_{2,4}$ (shown in Fig. 2.8) that $d(s) = 8 = h(G)$ or $d(s) = 10$ for *every* cyclic ordering s of $V(G)$.

For a connected graph G in general, the *upper Hamiltonian number* $h^+(G)$ is defined as

$$h^+(G) = \max \{d(s)\},$$

where the maximum is taken over all cyclic orderings s of the vertices of G. This concept was introduced in [23]. Thus, $h(G) = 8$ while $h^+(G) = 10$ for $G = K_{2,4}$. In fact, Theorem 2.39 implies that both $h(G)$ and $h^+(G)$ are even when G is bipartite.

Obviously, $h^+(G) \geq h(G)$ for every connected graph G in general, while the two parameters are equal when G is complete. As another example, let us consider the hypercubes Q_n. Note that $Q_1 = K_2$ and so $h(Q_1) = h^+(Q_1) = 2$. For $n \geq 2$, the graph Q_n is Hamiltonian and so $h(Q_n) = 2^n$ for each $n \geq 1$. The upper Hamiltonian number of Q_n was obtained in [23].

Theorem 2.43. *For each integer $n \geq 2$, $h^+(Q_n) = 2^{n-1}(2n - 1)$.*

Proof. First, we show that $h^+(Q_n) \leq 2^{n-1}(2n - 1)$. Let s be an arbitrary cyclic ordering of $V(Q_n)$ with $d(s) = h^+(Q_n)$. Since $\text{diam}(Q_n) = n$ and each vertex $v \in V(Q_n)$ has exactly one vertex $v' \in V(Q_n)$ such that $d(v, v') = n$, at most 2^{n-1} terms in $d(s)$ are equal to n. Thus, $h^+(Q_n) = d(s) \leq 2^{n-1}n + 2^{n-1}(n - 1) = 2^{n-1}(2n - 1)$.

To verify that $h^+(Q_n) \geq 2^{n-1}(2n - 1)$, note that the result is straightforward to verify for $Q_2 = C_4$ and so we may assume that $n \geq 3$. Let $G = Q_n$. Then G consists of two disjoint copies G_1 and G_2 of Q_{n-1}, where corresponding vertices of G_1 and G_2 are adjacent. For each vertex v of G, there is a unique vertex v' of G

such that $d(v, v') = n = \text{diam}(Q_n)$. Necessarily, exactly one of v and v' belongs to G_1. Let $(v_1, v_2, \ldots, v_{2^{n-1}}, v_{2^{n-1}+1} = v_1)$ be a Hamiltonian cycle in G_1 and consider the cyclic ordering s : $v_1, v'_1, v_2, v'_2, \ldots, v_{2^{n-1}}, v'_{2^{n-1}}, v_1$ of $V(G)$. By the triangle inequality, $d(v_{i+1}, v'_i) \geq d(v_i, v'_i) - d(v_i, v_{i+1}) = n - 1$ for $1 \leq i \leq 2^{n-1}$. Hence, $h^+(Q_n) \geq d(s) = 2^{n-1}n + 2^{n-1}(n-1) = 2^{n-1}(2n-1)$.   □

The upper Hamiltonian numbers of trees and cycles have been calculated in [23, 47].

Theorem 2.44. *If T is a nontrivial tree of order n, then*

$$2(n-1) = h(T) \leq h^+(T) \leq \lfloor n^2/2 \rfloor .$$

Furthermore, $h^+(T) = 2(n-1)$ if and only if T is a star and $h^+(T) = \lfloor n^2/2 \rfloor$ if and only if T is a path.

Theorem 2.45. *For each integer $n \geq 3$, $h(C_n) = n$ and*

$$h^+(C_n) = (n-2)\lfloor (n-1)/2 \rfloor + 2\lceil (n-1)/2 \rceil.$$

Theorems 2.43–2.45 show, not surprisingly, that $h^+(G)$ can be considerably larger than $h(G)$. There are, however, only two graphs G of a fixed order for which $h(G) = h^+(G)$, a fact established in [23]. Two vertices u and v are *antipodal vertices* in a connected graph G if $d(u, v) = \text{diam}(G)$.

Theorem 2.46. *Let G be a nontrivial connected graph. Then $h(G) = h^+(G)$ if and only if G is either complete or a star.*

Proof. Let G be a connected graph of order $n \geq 2$. For every cyclic ordering s of $V(G)$, observe that $d(s) = n$ if G is complete while $d(s) = 2(n-1)$ if G is a star. In other words, $h(K_n) = h^+(K_n) = n$ and $h(K_{1,n-1}) = h^+(K_{1,n-1}) = 2(n-1)$.

For the converse, suppose that G is a connected graph of order n and $G \neq K_n, K_{1,n-1}$. Thus, $n \geq 4$ and $d = \text{diam}(G) \geq 2$. We may also assume by Theorems 2.44 and 2.45 that G is neither a path nor a cycle. We now consider two cases, according to whether $d \geq 3$ or $d = 2$.

Case 1. $d \geq 3$. Let $P = (v_1, v_2, \ldots, v_{d+1})$ be a $v_1 - v_{d+1}$ geodesic, where v_1 and v_{d+1} are antipodal vertices in G. Since G itself is not a path, the set $U = V(G) - V(P)$ is not empty. Write $U = \{u_1, u_2, \ldots, u_{n-d-1}\}$ and define cyclic orderings s and s' of $V(G)$ by

$$s : v_1, v_2, v_3, v_4, \ldots, v_{d+1}, u_1, u_2, \ldots, u_{n-d-1}, v_1 \qquad (2.3)$$

$$s' : v_1, v_3, v_2, v_4, \ldots, v_{d+1}, u_1, u_2, \ldots, u_{n-d-1}, v_1. \qquad (2.4)$$

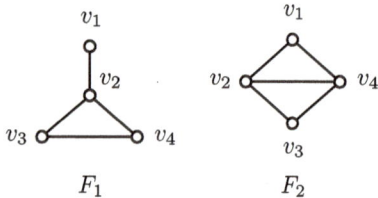

Fig. 2.9 Induced subgraphs F_1 and F_2 of G

Observe then that $h^+(G) \geq d(s') = d(s) + 2 \geq h(G) + 2$.

Case 2. $d = 2$. Since G is not a star, G contains cycles. Let g be the *girth* (the length of a smallest cycle) in G. So $g \geq 3$. Assume first that $g = 3$. Since G is connected and not complete, there exists a set $V \subseteq V(G)$ such that the subgraph induced by V in G is isomorphic to one of the graphs F_1 and F_2 in Fig. 2.9.

If $n = 4$, then the cyclic orderings $s : v_1, v_2, v_3, v_4, v_1$ and $s' : v_1, v_3, v_2, v_4, v_1$ show that $h^+(G) \geq h(G) + 1$. For $n \geq 5$, the set $U = V(G) - V$ is nonempty. Then define the cyclic orderings s and s' of $V(G)$ as described in (2.3) and (2.4) with $d = 3$, respectively, and verify that $d(s') = d(s) + 1$. Thus, $h^+(G) \geq h(G) + 1$.

If $g \geq 4$, then let $C = (v_1, v_2, \ldots, v_g, v_1)$ be an induced cycle of G and let $U = V(G) - V(C) = \{u_1, u_2, \ldots, u_{n-g}\}$, which is nonempty since $G \neq C_n$. Again, by considering the cyclic orderings s and s' of $V(G)$ as described in (2.3) and (2.4) with $d = g - 1$, respectively, we see that $h^+(G) \geq h(G) + 2$. $\qquad\square$

The proof of Theorem 2.46 suggests that if G is a graph with $h^+(G) - h(G) = 1$, then $|V(G)| \geq 4$, $\mathrm{diam}(G) = 2$, and G must contain a triangle. In order to obtain a complete characterization of those graphs G for which the difference between $h(G)$ and $h^+(G)$ is exactly 1, the following is useful. Note that $G \vee H$ denotes the join of vertex-disjoint graphs G and H (while $G + H$ is the union of G and H).

Lemma 2.2. *For a graph G, let $G_1 = K_1 \vee G$ and $G_2 = K_1 \vee \overline{G}$. Then $h^+(G_1) - h(G_1) = h^+(G_2) - h(G_2)$.*

Proof. For a graph G of order $n - 1$ (≥ 1), construct each of G_1 and G_2 by adding a new vertex and joining it to every vertex of G and $\overline{G}$, respectively. Let $V = V(G_1) = V(G_2)$. Since $G_1 = G_2 = K_2$ if $n = 2$, we may assume that $n \geq 3$. For every two distinct vertices $u, v \in V(G)$, we have $d_{G_1}(u, v) + d_{G_2}(u, v) = 3$. Therefore, $d_{G_1}(s) + d_{G_2}(s) = 3n - 2$ for every cyclic ordering s of V. Let s_1 and s_2 be cyclic orderings of $V(G_1) = V(G_2)$ such that $d_{G_1}(s_1) = h(G_1)$ and $d_{G_2}(s_2) = h^+(G_2)$. Then $3n - 2 = d_{G_1}(s_1) + d_{G_2}(s_1) \leq h(G_1) + h^+(G_2) \leq d_{G_1}(s_2) + d_{G_2}(s_2) = 3n - 2$, implying that $h(G_1) + h^+(G_2) = 3n - 2$. Similarly, $h(G_2) + h^+(G_1) = 3n - 2$. Therefore, $h^+(G_1) - h(G_1) = h^+(G_2) - h(G_2)$. $\qquad\square$

For a set $U \subseteq V(G)$, where say $|U| = \ell$, an ordering $v_1, v_2, \ldots, v_\ell$ of the ℓ vertices in U is called a *linear ordering* of U.

Theorem 2.47. *Let G be a nontrivial connected graph of order n. Then $h^+(G) - h(G) = 1$ if and only if $n \geq 4$ and $G = K_1 \vee H$, where*

$$H \in \{K_{1,\dots,1,2}, \overline{K_{1,\dots,1,2}}, K_{1,n-2}, \overline{K_{1,n-2}}\}.$$

Proof. For $n \geq 4$, let $H_1 = \overline{K_{1,n-2}}$, $H_2 = K_{1,\dots,1,2}$, $H_3 = \overline{H_2}$, and $H_4 = \overline{H_1}$. Then it is straightforward to verify that

$$h(K_1 \vee H_i) = h^+(K_1 \vee H_i) - 1 = \begin{cases} n+2-i & \text{if } i = 1,2 \\ 2n-i & \text{if } i = 3,4. \end{cases}$$

For the converse, suppose that G is a connected graph of order n and $h^+(G) - h(G) = 1$. Then $n \geq 4$ since G is neither complete nor a star by Theorem 2.46. Furthermore, as we saw in the proof of Theorem 2.46, there is neither P_4 nor C_4 as an induced subgraph in G. We may therefore assume that $\Delta(G) = n-1$ and G contains triangles. That is, $G = K_1 \vee H$ for some graph H of order $n-1$ that is neither complete nor empty. For $n = 4$, therefore, $H \in \{K_{1,2}, \overline{K_{1,2}}\}$.

Now assume that $n \geq 5$. We next show that none of $2K_2$, P_4, and C_4 is an induced subgraph in H. We have already seen that neither P_4 nor C_4 can be an induced subgraph in G, that is, neither is contained in H as an induced subgraph. Also, $2K_2 = \overline{C}_4$ cannot be an induced subgraph in H by Lemma 2.2. For $n = 5$, therefore, $H \in \{K_{1,1,2}, \overline{K_{1,1,2}}, K_{1,3}, \overline{K_{1,3}}\}$ or $H \in \{H_0, \overline{H_0}\}$, where $H_0 = K_1 + P_3$. One can quickly verify that $h(K_1 \vee H_0) = 6 = h^+(K_1 \vee H_0) - 2$ and so $h^+(K_1 \vee H_0) - h(K_1 \vee H_0) = h^+(K_1 \vee \overline{H_0}) - h(K_1 \vee \overline{H_0}) = 2$ by Lemma 2.2.

Finally, assume that $n \geq 6$. We next show that $\deg_H v \in \{0, 1, n-3, n-2\}$ for every $v \in V(H)$. Assume, to the contrary, that v_1 is a vertex in H with $2 \leq \deg_H v_1 \leq n-4$. Then let v_2, v_3, v_4, v_5 be vertices in H such that v_2 and v_3 are adjacent to v_1 while v_4 and v_5 are not. Let v_0 be the vertex in G that is adjacent to every vertex in H. Then by considering two orderings $s_1 : v_2, v_1, v_3, v_4, v_0, v_5, v_2$ and $s_2 : v_2, v_0, v_3, v_4, v_1, v_5, v_2$ (and by inserting some fixed linear ordering of $V(G) - \{v_0, v_1, \dots, v_5\}$ between v_5 and v_2 in each of s_1 and s_2 in case $n \geq 7$), we see that $h^+(G) - h(G) \geq 2$. This verifies the claim. Furthermore, $\Delta(H) \in \{1, n-3, n-2\}$ since H is nonempty. If $\Delta(H) = 1$, then $H = \overline{K_{1,\dots,1,2}}$ since $2K_2$ cannot be an induced subgraph in H. Thus, we now consider the following two cases. Let $V(H) = \{v_1, v_2, \dots, v_{n-1}\}$ and $\deg_H v_1 = \Delta(H)$.

Case 1. $\Delta(H) = n-3$. Then suppose that $v_1 v_2 \notin E(H)$. If $\deg_H v_2 \geq 1$, say $v_2 v_3 \in E(H)$, then we may assume that $v_3 v_4 \notin E(H)$ since $\deg_H v_3 \leq n-3$. However, this implies that the subgraph induced by $\{v_1, v_2, v_3, v_4\}$ is either C_4 or P_4, which cannot occur. Hence, $\deg_H v_2 = 0$. If $H \neq K_{1,n-2}$, then $H = K_1 + K_{1,n-3}$ since $\deg_H v \in \{0, 1, n-3, n-2\}$ for every $v \in V(H)$. To see that this cannot occur, observe that $\overline{H}$ is traceable and $K_1 \vee \overline{H}$ is Hamiltonian while $d_{K_1 \vee \overline{H}}(s) \geq n+2$ for any cyclic ordering of $V(K_1 \vee \overline{H})$ whose first three terms are v_3, v_1, v_4. Thus,

$h^+(K_1 \vee H) - h(K_1 \vee H) = h^+(K_1 \vee \overline{H}) - h(K_1 \vee \overline{H}) \geq 2$ by Lemma 2.2. Therefore, $H = \overline{K_{1,n-2}}$ is the only possibility in this case.

Case 2. $\Delta(H) = n-2$. Then $\delta(H) \in \{1, n-3\}$ since H is not complete. If there are two or more vertices having degree $n - 2$ in H, then $\delta(H) = n - 3$. Furthermore, $H = \overline{K_{1,\ldots,1,2}}$ since C_4 cannot occur as an induced subgraph in H. On the other hand, if v_1 is the only vertex whose degree in H equals $n - 2$, then the number of end-vertices in H is either 1 or $n - 2$. If the former occurs, then $\overline{H} = K_1 + K_{1,n-3}$. However, this is impossible by Case 1 and Lemma 2.2. Therefore, $H = \overline{K_{1,n-2}}$. □

Observe that if $s : v_1, v_2, \ldots, v_n, v_{n+1} = v_1$ is any cyclic ordering of the vertices of a connected graph, then for each vertex v_i $(1 \leq i \leq n)$, both $d(v_{i-1}, v_i) \leq e(v_i)$ and $d(v_i, v_{i+1}) \leq e(v_i)$, where $e(v_i)$ denotes the *eccentricity* of v_i (the distance from v_i to a vertex farthest from v_i). Therefore, if G is a connected graph of order $n \geq 3$ with $V(G) = \{v_1, v_2, \ldots, v_n\}$, then

$$h^+(G) \leq \sum_{i=1}^{n} e(v_i).$$

Since the eccentricity of a vertex in G is at most the diameter of G, we have the following upper bound for $h^+(G)$ in terms of the order and diameter of G.

Theorem 2.48. *If G is a nontrivial connected graph of order n and diameter d, then $h^+(G) \leq nd$.*

The upper bound in Theorem 2.48 has been shown to be sharp in [23]. A sharp lower bound for the upper Hamiltonian number of a connected graph G, also in terms of the order and diameter of G, was obtained by Král, Tong, and Zhu in [47].

Theorem 2.49. *If G is a nontrivial connected graph of order n and diameter d, then*

$$h^+(G) \geq n + \lceil d^2/2 \rceil - 1.$$

Furthermore, for each pair n, d of integers satisfying $1 \leq d \leq n - 1$, there is a connected graph G of order n and diameter d with $h^+(G) = n + \lceil d^2/2 \rceil - 1.$

2.8   The Hamiltonian Spectrum of a Graph

For a connected graph G, the *Hamiltonian spectrum* $\mathscr{H}(G)$ of G is defined in [47] as

$$\mathscr{H}(G) = \{d(s) : s \text{ is a cyclic ordering of the vertices of } G\}.$$

Fig. 2.10 Illustrating the
Hamiltonian spectrum of a
graph

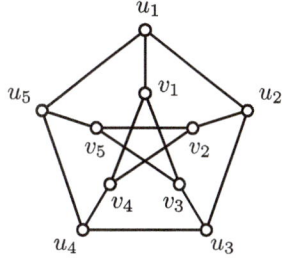

Of course, this implies that $h(G), h^+(G) \in \mathcal{H}(G)$ and, in general,

$$\mathcal{H}(G) \subseteq \{k : k = h(G), h(G) + 1, \ldots, h^+(G)\}. \tag{2.5}$$

The following is therefore an immediate consequence of Theorem 2.46.

Theorem 2.50. *The Hamiltonian spectrum of a connected graph G consists of a single number if and only if G is either a complete graph or a star.*

As another illustration, consider the Petersen graph P in Fig. 2.10. Since P is a non-Hamiltonian graph of order 10, it follows that $h(P) \geq 11$. On the other hand, $h^+(P) \leq 20$ by Theorem 2.48. Therefore, $11 \leq h(P) < h^+(P) \leq 20$. In fact, $h(P) = 11$ and $h^+(P) = 20$. Consider the sequences s_i $(1 \leq i \leq 10)$ given by

$$
\begin{aligned}
s_1 &: u_1, u_2, u_3, u_4, u_5, v_5, v_2, v_4, v_3, v_1, u_1 \\
s_2 &: u_1, u_2, u_3, u_4, u_5, v_5, v_2, v_3, v_4, v_1, u_1 \\
s_3 &: u_1, u_2, u_3, u_5, u_4, v_4, v_2, v_3, v_5, v_1, u_1 \\
s_4 &: u_1, u_3, u_5, u_2, u_4, v_4, v_2, v_5, v_3, v_1, u_1 \\
s_5 &: u_1, u_3, u_5, u_2, u_4, v_3, v_5, v_2, v_4, v_1, u_1 \\
s_6 &: u_1, u_3, u_5, u_2, u_4, v_5, v_2, v_4, v_3, v_1, u_1 \\
s_7 &: u_1, u_3, u_5, u_2, u_4, v_3, v_5, v_4, v_2, v_1, u_1 \\
s_8 &: u_1, u_3, u_5, u_2, v_2, u_4, v_3, v_4, v_5, v_1, u_1 \\
s_9 &: u_1, u_3, u_5, u_2, u_4, v_2, v_3, v_4, v_5, v_1, u_1 \\
s_{10} &: u_1, u_3, u_5, u_2, u_4, v_1, v_2, v_3, v_4, v_5, u_1.
\end{aligned}
$$

Since $d(s_i) = 10 + i$ for $1 \leq i \leq 10$, it follows that $\mathcal{H}(P) = \{11, 12, \ldots, 20\}$, that is, equality holds in (2.5). On the other hand, Theorem 2.39 implies that the Hamiltonian spectrum of a connected bipartite graph consists only of even integers, that is, equality in (2.5) does not hold in general.

The Hamiltonian spectrum of an n-cycle was determined in [47] for each integer $n \geq 3$. Recall that $h^+(C_n) = (n - 2)\lfloor (n - 1)/2 \rfloor + 2\lceil (n - 1)/2 \rceil$.

Theorem 2.51. *Let $n \geq 3$ be an integer.*
(a) *If n is even, then $\mathcal{H}(C_n) = \{n, n + 2, \ldots, h^+(C_n) - 2, h^+(C_n)\}$.*

(b) *If n is odd, then*

$$\mathscr{H}(C_n) = \{n, n+2, \ldots, 2n-5, 2n-3\} \cup$$
$$\{2n-2, 2n-1, \ldots, h^+(C_n) - 2\} \cup \{h^+(C_n)\}.$$

The Hamiltonian spectrum of a tree was determined by Liu [50]. In order to present this result, we introduce some additional definitions. For a vertex v of a connected graph G, the *total distance* $\mathrm{td}(v)$ of v is the sum of the distances from v to all vertices of G. The minimum total distance over all vertices of G is the *median number* of G and is denoted by $\mathrm{med}(G)$.

Theorem 2.52. *For a nontrivial tree T of order n,*

$$\mathscr{H}(T) = \{2k : k = n-1, n, n+1, \ldots, \mathrm{med}(T)\}.$$

The following is a consequence of Theorem 2.52.

Theorem 2.53. *The upper Hamiltonian number of a nontrivial tree T equals $2\,\mathrm{med}(T)$.*

According to Theorems 2.44 and 2.53 (or Theorem 2.39), the upper Hamiltonian number of a tree of order n is an even integer between $2(n-1)$ and $\lfloor n^2/2 \rfloor$. In fact, for each integer $n \geq 3$, every even integer between $2(n-1)$ and $\lfloor n^2/2 \rfloor$ is the upper Hamiltonian number of some tree of order n. In order to show this, we first present some preliminary results. A vertex of a connected graph G whose total distance equals the median number of G is a *median vertex* of G. The subgraph of G induced by its median vertices of G is the *median* of G. The following two lemmas will be useful to us, the first of which is an easy observation and the second of which was established by Truszczyński [65].

Lemma 2.3. *No end-vertex of a tree T of order at least 3 is a median vertex of T.*

Lemma 2.4. *The median of every connected graph G lies in a single block of G.*

It therefore follows by Lemma 2.4 that the median of a tree is either K_1 or K_2.

Theorem 2.54. *For each pair n, k of integers satisfying $1 \leq n-1 \leq k \leq \lfloor n^2/4 \rfloor$, there exists a tree T of order n such that $h^+(T) = 2k$.*

Proof. By Theorem 2.44, the result holds when $k \in \{n-1, \lfloor n^2/4 \rfloor\}$. Thus, let $n \geq 5$ be a fixed integer and suppose that k is an integer satisfying $n+1 \leq k \leq \lfloor n^2/4 \rfloor$ and there exists a tree T_k of order n with $h^+(T_k) = 2k$. We show that there exists a tree T of order n with $h^+(T) = 2(k-1)$.

Let x be a median vertex of T_k and select a vertex y farthest from x. Thus, y is an end-vertex in T_k while x is not by Lemma 2.3. Also, $\text{td}_{T_k}(x) = \text{med}(T_k) = k$ by Theorem 2.53. Now consider the $y - x$ geodesic $P = (y = v_0, v_1, v_2, \ldots, v_{e(x)} = x)$, where $e(x)$ is the eccentricity of x. Note that $e(x) \geq 2$ since T_k is not a star. Let T be the tree obtained from T_k by deleting the edge $v_0 v_1$ and adding the edge $v_0 v_2$. Then y is an end-vertex in T while v_1 may or may not be. We claim that $\text{med}(T) = k - 1$. For each vertex $v \in V(T) - \{y\}$, observe that

$$\text{td}_T(v) = \begin{cases} \text{td}_{T_k}(v) + 1 & \text{if } v \in V(T') \\ \text{td}_{T_k}(v) - 1 & \text{otherwise,} \end{cases}$$

where T' is the component of $T - v_1 v_2$ containing v_1. Since $\text{td}_T(y) > \text{med}(T)$ again by Lemma 2.3, it follows that $\text{med}(T) = \text{td}_T(x) = k - 1$ and so $h^+(T) = 2(k-1)$ by Theorem 2.53.   □

As we have seen earlier, $d(s)$ and $d(s')$ are of the same parity for every two cyclic orderings s and s' of $V(G)$ if G is either complete or bipartite. In fact, these are the only two classes of connected graphs with this property.

Theorem 2.55 ([35]). *A nontrivial connected graph G has the property that $d(s)$ and $d(s')$ are of the same parity for every two cyclic orderings s and s' of $V(G)$ if and only if G is complete or bipartite.*

Proof. By the discussion above, we may assume that G is neither complete nor bipartite. We consider two cases.

Case 1. G contains a triangle. Let $G' = K_\omega$ be a largest clique in G, where then $\omega \geq 3$. Since G is not complete and G is connected, there is a vertex in $V(G) - V(G')$ that is adjacent to some but not all vertices of G'. Thus, there is a triangle (v_1, v_2, v_3, v_1) and a vertex $v_4 \in V(G) - \{v_1, v_2, v_3\}$ such that $v_2 v_4 \notin E(G)$ and $v_3 v_4 \in E(G)$. For a fixed linear ordering s of $V(G) - \{v_2, v_3, v_4\}$ whose terminal vertex is v_1, let s_1 be the ordering v_1, v_2, v_3, v_4 followed by s. Similarly, let s_2 be the ordering v_1, v_3, v_2, v_4 followed by s. Then both s_1 and s_2 are cyclic orderings of $V(G)$ and $d(s_2) - d(s_1) = 1$. Hence, $d(s_1)$ and $d(s_2)$ are of opposite parity.

Case 2. G is triangle-free. Let $C = (v_1, v_2, \ldots, v_\ell, v_1)$ be a shortest odd cycle in G. Thus, $\ell \geq 5$ and C is an induced subgraph of G. We consider two subcases.

Subcase 2.1. $\ell = 5$. If G itself is a cycle, that is, if $n = \ell = 5$, then let s_1 : $v_1, v_3, v_2, v_4, v_5, v_1$ and s_2 : $v_1, v_3, v_4, v_2, v_5, v_1$ be two cyclic orderings of the vertices of G. Hence, $d(s_1) = 7$ and $d(s_2) = 8$. If $n \geq 6$, then let s be a fixed linear ordering of the vertices of $V(G) - \{v_2, v_3, v_4, v_5\}$ whose terminal vertex is v_1. Now consider s_1' : v_1, v_3, v_2, v_4, v_5 and s_2' : v_1, v_4, v_2, v_3, v_5. For $i = 1, 2$, let s_i be the ordering s_i' followed by s. Then s_i is a cyclic ordering of $V(G)$ and $d(s_2) - d(s_1) = 1$.

Subcase 2.2. $\ell \geq 7$. Let s be a fixed linear ordering of the set $V(G) - \{v_2, \ldots, v_{\ell-1}\}$ whose terminal vertex is v_1. Let $\ell^* = (\ell + 1)/2$ and consider

$$s_1' : v_1, v_{\ell^*}, v_2, v_3, \ldots, v_{\ell^*-1}, v_{\ell^*+1}, v_{\ell^*+2}, \ldots, v_{\ell-1}$$

$$s_2' : v_1, v_{\ell^*+1}, v_2, v_3, \ldots, v_{\ell^*}, v_{\ell^*+2}, v_{\ell^*+3}, \ldots, v_{\ell-1}.$$

For $i = 1, 2$, let s_i be the ordering s_i' followed by s. Then both s_1 and s_2 are cyclic orderings of $V(G)$ and $d(s_2) - d(s_1) = 1$. □

The following is an immediate consequence of the proof of Theorem 2.55.

Theorem 2.56. *If G is a nontrivial connected graph that is neither complete nor bipartite, then there are cyclic orderings s and s' of $V(G)$ such that $d(s) - d(s') = 1$. In other words, $\mathcal{H}(G)$ contains two consecutive integers.*

We have seen that the Hamiltonian spectrum of a graph G consists of a single element if and only if G is either complete or a star. Suppose now that G is a graph for which $\mathcal{H}(G)$ contains exactly two elements. If G is not bipartite, then it follows by Theorem 2.56 that $h^+(G) - h(G) = 1$. Such graphs have been completely characterized in Theorem 2.47.

A tree T is a *double star* if it contains exactly two vertices that are not end-vertices. Necessarily, these two vertices are adjacent in T. If their degrees are r and s $(r, s \geq 2)$, respectively, then we write $T = S_{r,s}$. For those graphs G that are bipartite and $|\mathcal{H}(G)| = 2$, we have the following.

Theorem 2.57 ([35]). *Let G be a nontrivial connected bipartite graph of order n. Then $|\mathcal{H}(G)| = 2$ if and only if $n \geq 4$ and G is either $S_{2,n-2}$ or $K_{2,n-2}$.*

Combining Theorems 2.56 and 2.57, we have the following.

Theorem 2.58. *Let G be a nontrivial connected graph of order n. Then $|\mathcal{H}(G)| = 2$ if and only if $n \geq 4$ and either*
(a) *$G \in \{S_{2,n-2}, K_{2,n-2}\}$ or*
(b) *$G = K_1 \vee H$, where $H \in \{K_{1,\ldots,1,2}, \overline{K_{1,\ldots,1,2}}, K_{1,n-2}, \overline{K_{1,n-2}}\}$.*
Furthermore, the two integers in $\mathcal{H}(G)$ are of the same parity if and only if (a) occurs.

Theorem 2.59 ([35]). *If G is a connected graph of order n such that $h^+(G) - h(G) = 2$, then exactly one of the following (a)–(c) occurs:*
(a) *$n \geq 4$ and $G \in \{S_{2,n-2}, K_{2,n-2}\}$.*
(b) *$n \geq 5$ and $G = H_1 \vee H_2$, where H_1 is complete and*
    i. *$n \geq 6$ and $H_2 = \overline{K}_3$ or*
    ii. *$n \geq 5$ and $H_2 = K_{2,2}$ or*
    iii. *$n \geq 5$ and $H_2 = K_1 + K_\ell$, where $2 \leq \ell \leq n - 3$.*
(c) *$n \geq 5$ and G is neither bipartite nor Hamiltonian.*

Traceable Walks

3

3.1 The Traceable Number of a Graph

Recall that a graph is Hamiltonian if it contains a Hamiltonian cycle and a graph is traceable if it contains a Hamiltonian path. Therefore, every Hamiltonian graph is traceable but the converse is clearly false.

As a corollary of some results by Dirac and Ore (Theorems 2.1 and 2.2), we saw some sufficient conditions for a graph to be traceable (Theorem 2.3). For example, a graph G is traceable provided its minimum degree is at least $(|V(G)| - 1)/2$.

For a graph G, let S be a nonempty proper subset of $V(G)$. We have seen that

$$k(G - S) \le \begin{cases} |S| & \text{if } G \text{ is Hamiltonian} \\ |S| - 1 & \text{if } G \text{ is Hamiltonian-connected,} \end{cases}$$

where $k(H)$ denotes the number of components in a graph H (Theorems 2.11 and 2.14). There is a similar necessary condition for graphs to be traceable.

Theorem 3.1. *If G is traceable, then $k(G - S) \le |S| + 1$ for every nonempty proper subset S of $V(G)$.*

Proof. Suppose that G contains a Hamiltonian path and consider the Hamiltonian graph H obtained from G by adding a new vertex v and joining it to every vertex in G. If S is a nonempty proper subset of $V(G)$, then consider the set $S' = S \cup \{v\}$. Since H is Hamiltonian and S' is a nonempty proper subset of $V(H)$, it follows by Theorem 2.11 that $k(G - S) = k(H - S') \le |S'| = |S| + 1$. □

We saw in Sect. 2.6 that a Hamiltonian graph G of order n is a graph for which there is a cyclic ordering $v_1, v_2, \ldots, v_n, v_{n+1} = v_1$ of $V(G)$ such that $d(v_i, v_{i+1}) = 1$ for $1 \le i \le n$. We may view traceable graphs in a similar way. That is, if a nontrivial graph G of order n is traceable, then there is a *linear ordering* $v_1, v_2, \ldots, v_n$ of its vertices such that $d(v_i, v_{i+1}) = 1$ for $1 \le i \le n - 1$. We now define a new graphical parameter, which corresponds to the Hamiltonian number of graphs.

F. Fujie and P. Zhang, *Covering Walks in Graphs*, SpringerBriefs in Mathematics, DOI 10.1007/978-1-4939-0305-4_3, © Futaba Fujie, Ping Zhang 2014

For a connected graph G of order $n \geq 2$ and a linear ordering $s : v_1, v_2, \ldots, v_n$ of the vertices of G, we define the number $d(s)$ by

$$d(s) = \sum_{i=1}^{n-1} d(v_i, v_{i+1}).$$

The traceable number $t(G)$ of G is then defined by

$$t(G) = \min\{d(s)\},$$

where the minimum is taken over all linear orderings s of $V(G)$. Thus, $t(G) \geq n-1$ in general and $t(G) = n - 1$ if and only if G is traceable.

While a Hamiltonian walk in a graph G is a closed spanning walk of minimum length in G, an open spanning walk of minimum length in G is called a *traceable walk*. Hence, a traceable walk in G is a vertex-covering walk of minimum length. As the Hamiltonian number equals the length of a Hamiltonian walk for a graph, the traceable number of a graph and the length of a traceable walk in that graph are equal.

Theorem 3.2 ([53]). *For a nontrivial connected graph G, the length of a traceable walk in G equals $t(G)$.*

Several bounds for the traceable number of a graph have been established in terms of its order and other graphical parameters. A closed spanning walk W in a connected graph G always contains an open spanning walk in G whose length is at most $L(W) - 1$. Thus, if G is a connected graph of order n and size m, then

$$n - 1 \leq t(G) < h(G) \leq e(G) \leq 2m.$$

In fact, the difference $h(G) - t(G)$ is positive and bounded above by the diameter of G.

Theorem 3.3 ([53]). *If G is a nontrivial connected graph, then*

$$1 \leq h(G) - t(G) \leq \text{diam}(G).$$

Proof. The lower bound is immediate. For the upper bound, let $s : v_1, v_2, \ldots, v_n$ be a linear ordering of $V(G)$, where n is the order of G, such that $d(s) = t(G)$. Then for the cyclic ordering $s_c : v_1, v_2, \ldots, v_n, v_1$ of $V(G)$, observe that $h(G) \leq d(s_c) = d(s) + d(v_1, v_n) \leq t(G) + \text{diam}(G)$. Thus, $h(G) - t(G) \leq \text{diam}(G)$. □

Since $n \leq h(G) \leq 2(n - 1)$ (Theorem 2.25), we obtain the following bounds for the traceable number of a graph in terms of its order.

Theorem 3.4. *For every connected graph G of order n $\geq$ 2,*

$$n - 1 \leq t(G) \leq 2n - 3.$$

Obviously, the lower bound in Theorem 3.4 is sharp. For the upper bound, we will see soon that there is only one graph of order n whose traceable number equals $2n - 3$.

While the standard distance $d(u, v)$ between two vertices u and v in a connected graph G is the length of a $u - v$ geodesic (a shortest $u - v$ path) in G, this is not the only way that distance has been defined on the vertex set of a connected graph. The length of a longest $u - v$ path is called the *detour distance* $D(u, v)$ between u and v in G. As with standard distance, detour distance is a metric on $V(G)$. A $u - v$ path of length $D(u, v)$ is a $u - v$ *detour*. The *eccentricity* $e(v)$ of a vertex v in G is the distance from v to a vertex farthest from v in G. The *detour eccentricity* $e_D(v)$ of a vertex v is, as expected, the detour distance from v to a vertex farthest from v. The *detour diameter* $\operatorname{diam}_D(G)$ of G is then a maximum detour eccentricity among the vertices in G. In other words, $\operatorname{diam}_D(G)$ is the length of a longest path in G.

Theorem 3.5 ([53]). *If G is a nontrivial connected graph of order n, then*

$$t(G) \leq 2(n - 1) - \operatorname{diam}_D(G).$$

Proof. We proceed by induction on n. The result is straightforward to verify for $n = 2, 3$. For an integer $n \geq 4$, assume, for every connected graph H of order $n - 1$, that $t(H) \leq 2n - 4 - \operatorname{diam}_D(H)$. Let G be a connected graph of order n. If G is traceable, then $t(G) = \operatorname{diam}_D(G) = n - 1$; so $t(G) = 2(n - 1) - \operatorname{diam}_D(G)$ in this case. Hence, we assume that G does not contain a Hamiltonian path, that is, $\operatorname{diam}_D(G) \leq n - 2$. Let P be a $u - v$ detour of length $\operatorname{diam}_D(G)$ in G. Among the vertices in G not belonging to P, let w be a vertex such that the standard distance between w and a vertex in P is maximum. Thus, $G - w$ is a connected graph of order $n - 1$ containing P and so $\operatorname{diam}_D(G - w) = \operatorname{diam}_D(G)$. Let $s : v_1, v_2, \ldots, v_{n-1}$ be a linear ordering of $V(G - w)$ with $d(s) = t(G - w)$. Since G is connected, $wv_i \in E(G)$ for some i ($1 \leq i \leq n - 1$). Let s' be a linear ordering of $V(G)$ obtained from s by inserting w immediately after v_i. If $i = n$, then $t(G) \leq d(s') = d(s) + d(v_n, w) = t(G - w) + 1$. Otherwise, $d(w, v_{i+1}) \leq d(w, v_i) + d(v_i, v_{i+1})$ and so $t(G) \leq d(s') = d(s) + d(v_i, w) + d(w, v_{i+1}) - d(v_i, v_{i+1}) \leq t(G - w) + 2$. Since $t(G - w) \leq 2n - 4 - \operatorname{diam}_D(G - w) = 2n - 4 - \operatorname{diam}_D(G)$ by the induction hypothesis, the result now follows. $\qquad\square$

Theorem 3.5 implies that a graph G with $t(G) = 2|V(G)| - 3$ must have $\operatorname{diam}_D(G) = 1$, that is, $G = K_2$. Since $t(K_2) = 2 \cdot 2 - 3$, we now see that K_2 is the only graph that attains the upper bound in Theorem 3.4. In other words, $n - 1 \leq t(G) \leq 2n - 4$ for every connected graph G of order $n \geq 3$.

Fig. 3.1 A graph G with $t(G) = 13$

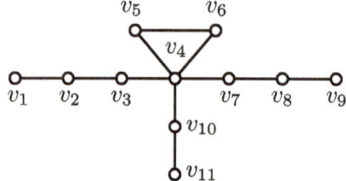

A graph is a *linear forest* if each of its components is a path. The following result gives a lower bound for the traceable number of a graph in terms of its order and the maximum size of a spanning linear forest.

Theorem 3.6 ([53]). *If G is a nontrivial connected graph of order n such that the maximum size of a spanning linear forest in G is p, then $t(G) \geq 2(n-1) - p$.*

Proof. For an arbitrary linear sequence $s : v_1, v_2, \ldots, v_n$ of $V(G)$, observe that at most p of the $n-1$ numbers $d(v_i, v_{i+1})$ $(1 \leq i \leq n-1)$ equal 1 and the remaining $n-1-p$ numbers are at least 2. Hence, $d(s) \geq p + 2(n-1-p) = 2(n-1) - p$, that is, $t(G) \geq 2(n-1) - p$. $\qquad\square$

By Theorems 3.5 and 3.6, for example, $t(K_{1,n-1}) = 2n - 4$ for $n \geq 3$. As another example, the graph G of Fig. 3.1 has order $n = 11$ and $\mathrm{diam}_D(G) = 6$ while the maximum size of a spanning linear forest in G is 8. Therefore, $12 \leq t(G) \leq 14$. It turns out that $t(G) = 13$ and $s : v_1, v_2, \ldots, v_{11}$ is a linear ordering of G with $d(s) = 13$. In order to show that $t(G) > 12$, consider a spanning linear forest F of G whose size equals 8. Then F consists of three paths, say Q_1, Q_2, and Q_3. Let Q_i be an $x_i - x_i'$ path for $1 \leq i \leq 3$. If the distance between two end-vertices belonging to different paths equals 2, say $d(x_1, x_2) = 2$, then $d(u, v) \geq 3$ for all $u \in \{x_1', x_2'\}$ and $v \in \{x_3, x_3'\}$. Now assume, to the contrary, that s' is a linear ordering of $V(G)$ with $d(s') = 12$. Then the ten summands in $d(s')$ are either (i) eight 1s and two 2s or (ii) nine 1s and one 3. By the above observation, (i) cannot occur. However, (ii) is also impossible since there is no spanning linear forest of size 9. Hence, no such s' exists and we conclude that $2(n-1) - p < t(G) < 2(n-1) - \mathrm{diam}_D(G)$ in this case.

The above observation gives us the following.

Theorem 3.7 ([53]). *If G is a nontrivial connected graph of order n and $\mathrm{diam}(G) = 2$, then $t(G) = 2(n-1) - p$, where p is the maximum size of a spanning linear forest in G.*

Suppose that v is a vertex in a connected graph G that is not a cut-vertex of G. Then $G - v$ is connected and $d_{G-v}(x, y) \geq d_G(x, y)$ for every $x, y \in V(G - v)$. The following result gives a bound for the traceable number of $G - v$ in terms of the traceable number of G and some other parameters.

Theorem 3.8. *Let v be a vertex in a nontrivial connected graph G such that G − v is connected. Then*

$$t(G) - 2 \le t(G - v) \le \begin{cases} t(G) - 1 & \text{if } v \text{ is an end-vertex or } \ell = 3 \\ t(G) + k(\ell - 4) & \text{if } \ell \ge 4, \end{cases}$$

where ℓ is the length of a longest cycle containing v in G (when v is not an end-vertex) and k is the minimum number of occurrences of v in a traceable walk of G.

Proof. First, let $W = (w_0, w_1, \ldots, w_{t(G-v)})$ be a traceable walk of $G - v$. Then let $w_i \in N(v)$ and consider the walk obtained from W by inserting the two vertices v, w_i immediately after w_i. Then the resulting walk is a spanning walk of G whose length equals $t(G - v) + 2$. Thus, $t(G) \le t(G - v) + 2$.

For the upper bound, let $W = (w_0, w_1, \ldots, w_{t(G)})$ be a traceable walk of G such that v occurs in W a total of k times. If v is an end-vertex or $\ell \in \{3, 4\}$, then $d_{G-v}(x, y) = d_G(x, y)$ for every $x, y \in V(G - v)$.

Case 1. v is an end-vertex or $\ell = 3$. Then the subgraph induced by $N(v)$ is complete and no $x - y$ geodesic in G contains v unless $v \in \{x, y\}$. Thus, v appears in W exactly once. If $v = w_0$ or $v = w_{t(G)}$, then we obtain a spanning walk of $G - v$ whose length equals $t(G) - 1$. Otherwise, we still have a spanning walk whose length equals (i) $t(G) - 2$ if v is an end-vertex and (ii) $t(G) - 1$ if $\ell = 3$. Hence, $t(G - v) \le t(G) - 1$ in this case.

Case 2. $\ell \ge 4$. If $v = w_i$ for some i $(1 \le i \le t(G) - 1)$, then $d_{G-v}(w_{i-1}, w_{i+1}) \le \ell - 2$. Thus, if $v \in \{w_0, w_{t(G)}\}$, then $t(G-v) \le t(G) - 1 - 2(k-1) + (\ell-2)(k-1) = t(G) + (k - 1)(\ell - 4) - 1$. Otherwise, $t(G - v) \le t(G) - 2k + (\ell - 2)k = t(G) + k(\ell - 4)$.

This completes the proof.   □

3.2   The Traceable Number of a Tree

Analogous to Hamiltonian numbers of graphs, the traceable number of a connected graph is bounded above by the traceable number of any connected spanning subgraph of that graph. In particular, if T is a spanning tree of G, then $t(G) \le t(T)$. For this reason, it is useful to study the traceable numbers of trees.

In a tree, every two vertices are connected by a unique path. Therefore, $d(u, v) = D(u, v)$ for every two vertices u, v in a tree, which then implies that $\operatorname{diam}(T) = \operatorname{diam}_D(T)$ for every tree T. The next result gives the exact value of the traceable number of a tree in terms of its order and diameter.

Theorem 3.9 ([53]). *If T is a nontrivial tree of order n, then*

$$t(T) = 2(n - 1) - \operatorname{diam}(T).$$

Proof. By Theorem 3.5, we have $t(G) \leq 2(n-1) - \mathrm{diam}_D(T) = 2(n-1) - \mathrm{diam}(T)$. On the other hand, we also have $t(T) \geq h(T) - \mathrm{diam}(T) = 2(n-1) - \mathrm{diam}(T)$ by Theorems 2.24 and 3.3.   □

We have seen that $n-1 \leq t(G) \leq 2n-4$ for a connected graph G of order $n \geq 3$. For a tree T of order $n \geq 3$, we have $2 \leq \mathrm{diam}(T) \leq n-1$ and $\mathrm{diam}(T) = 2$ if and only if T is a star while $\mathrm{diam}(T) = n-1$ if and only if T is a path. For an integer d with $3 \leq d \leq n-2$, let T be a tree obtained from a path P of order $d+1$ by adding $n-d-1$ new vertices and joining each of them to a vertex on P that is not an end-vertex. Since T is a tree of order n and $\mathrm{diam}(T) = d$, every pair n,d of integers with $2 \leq d \leq n-1$ is realizable as the order and diameter of a tree. This observation and Theorem 3.9 establish another realization result.

Theorem 3.10 ([53]). *For each pair k,n of integers with $2 \leq n-1 \leq k \leq 2n-4$, there exists a tree T of order n and $t(T) = k$.*

If T is a tree of order $n \geq 3$, then of course, $t(T) = n-1$ if and only if $T = P_n$. Also, $t(T) = 2n-4$ if and only if $T = K_{1,n-1}$. Thus, there is only one tree of order n having traceable number $n-1$ and there is only one tree of order n having traceable number $2n-4$.

Recall that a tree T is a double star if it contains exactly two vertices that are not end-vertices. Necessarily these vertices are adjacent in T. Equivalently, T is a double star if and only if $\mathrm{diam}(T) = 3$; that is, T is a tree of order n and traceable number $2n-5$ if and only if T is a double star. Hence, there are exactly $\lfloor n/2 \rfloor - 1$ non-isomorphic trees of order n having traceable number $2n-5$. Similarly, there are exactly $\lceil (n-3)/2 \rceil$ non-isomorphic trees of order n having traceable number n, since such trees must have diameter $n-2$, that is, each of these trees must be a caterpillar of order $n \geq 4$ containing exactly three end-vertices whose removal results in P_{n-3}.

Suppose that G is a nontrivial connected graph of order n. Then we have seen that $n-1 \leq t(G) \leq 2n-3$ and
- $t(G) = n-1$ if and only if G contains P_n as a spanning tree, that is, $\mathrm{diam}_D(G) = n-1$;
- $t(G) = 2n-3$ if and only if $G = K_2$, that is, $\mathrm{diam}_D(G) = 1$.

When G is not traceable, that is, if $n \leq t(G) \leq 2n-4$, then what can be said about G and its spanning trees? For graphs having some specific traceable numbers, we have the following:
(a) If $t(G) = 2n-4$, then G contains $K_{1,n-1}$ as a spanning tree but the converse does not hold.
(b) If $t(G) = 2n-5$, then G contains a double star of order n as a spanning tree but the converse does not hold.
(c) If $t(G) = n$, then G does not necessarily contain a caterpillar having diameter $n-2$ as a spanning tree.
We verify (a)–(c) by characterizing those graphs G with $t(G) \in \{n, 2n-5, 2n-4\}$, where n is the order of G.

Theorem 3.11. *If G is a nontrivial connected graph of order n, then $t(G) = 2n-4$ if and only if $\operatorname{diam}_D(G) = 2$.*

Proof. Since $t(K_2) > 2 \cdot 2 - 4$, it follows that $n \geq 3$ and $\operatorname{diam}_D(G) \geq 2$. Assume first that $t(G) = 2n - 4$. Then $\operatorname{diam}_D(G) \leq 2$ by Theorem 3.5. Conversely, if $\operatorname{diam}_D(G) = 2$, then $\Delta(G) = n - 1$ since otherwise a path whose length exceeds 2 results. Therefore, G is either a triangle or a star. Note that $t(K_3) = 2 \cdot 3 - 4$ while $t(K_{1,n-1}) = 2n - 4$ by Theorem 3.9. □

Theorem 3.12. *If G is a connected graph of order n, then $t(G) = 2n - 5$ if and only if $\operatorname{diam}_D(G) = 3$.*

Proof. By Theorem 3.11, we may assume that $\operatorname{diam}_D(G) \geq 3$. Also, $\operatorname{diam}_D(G) \leq 2(n-1) - t(G)$ by Theorem 3.5. Thus, $t(G) = 2n - 5$ implies that $\operatorname{diam}_D(G) = 3$. Conversely, suppose that $\operatorname{diam}_D(G) = 3$. By Theorem 3.5, it suffices to verify that $t(G) \geq 2n - 5$. Since the result immediately follows when G is a double star by Theorem 3.9, suppose that G contains a cycle. Then either (i) $n = 4$ and G is traceable or (ii) $n \geq 5$ and $G = K_{1,n-1} + e$. If (i) occurs, then $t(G) = 2 \cdot 4 - 5$. Otherwise, the maximum size of a spanning linear forest in G is 3 and so $t(G) \geq 2n - 5$ by Theorem 3.6. □

Theorems 3.11 and 3.12 imply the following.

Theorem 3.13. *Let G be a connected graph of order $n \geq 3$.*
(a) $t(G) = 2n - 4$ if and only if $G \in \{K_3, K_{1,n-1}\}$.
(b) $t(G) = 2n - 5$ if and only if either
     i. $n = 4$ and $G \neq K_{1,3}$ or
    ii. $n \geq 5$ and either G is a double star or $n = |E(G)| = \Delta(G) + 1$.

If T is a tree containing exactly three end-vertices, then $\Delta(T) = 3$ and there is exactly one vertex whose degree equals 3. In other words, T is a subdivision of a claw $K_{1,3}$. We next characterize those graphs G for which $t(G) = |V(G)|$.

Theorem 3.14. *If G is a connected graph of order $n \geq 3$, then $t(G) = n$ if and only if (i) $\operatorname{diam}_D(G) \leq n - 2$ and (ii) G contains a spanning tree T with exactly three end-vertices, one of which is adjacent in G to the vertex of degree 3 in T.*

Proof. Suppose that $t(G) = n$ and let $W = (w_0, w_1, \ldots, w_n)$ be a traceable walk in G. Since W contains $n + 1$ vertices, $w_i = w_j$ for some i, j with $0 \leq i < j \leq n$ and $j - i \geq 2$. We may also assume that $i \neq 0$ and $j \neq n$ since G is not traceable. Therefore, G contains a spanning tree T in which w_0, w_{j-1} and w_n are the only end-vertices in T and $\Delta(T) = \deg_T(w_i) = 3$. Furthermore, $w_i w_{j-1} \in E(G)$. For the converse, suppose that G contains a spanning tree T obtained from three nontrivial paths $(u_0, u_1, \ldots, u_{\ell_1})$, $(v_0, v_1, \ldots, v_{\ell_2})$, and $(w_0, w_1, \ldots, w_{\ell_3})$ by identifying the three vertices u_0, v_0, and w_0. Furthermore, suppose that u_0 and u_{ℓ_1} are adjacent in

G. Then $s : v_{\ell_2}, v_{\ell_2-1}, \ldots, v_1, u_0, u_1, \ldots, u_{\ell_1}, w_1, w_2 \ldots, w_{\ell_3}$ is a linear ordering of $V(G)$ and $d(s) \leq n$ since the only pair of consecutive vertices in s that are possibly not adjacent is u_{ℓ_1}, w_1 and $d(u_{\ell_1}, w_1) \leq 2$. If $\operatorname{diam}_D(G) \leq n - 2$, then G is not traceable and so $t(G) = n$. ☐

Thus far, we have seen that if G is a nontrivial connected graph of order n, then $n - 1 \leq t(G) \leq 2n - 3$ and $t(G) + \operatorname{diam}_D(G) = 2(n - 1)$ whenever $t(G) \in \{n - 1, 2n - 5, 2n - 4, 2n - 3\}$. However, if $t(G) = 2n - 6$, then $\operatorname{diam}_D(G) = 4$ by Theorems 3.5, 3.11, and 3.12 but the converse is false. To see this, let ℓ be a nonnegative integer and $G = K_1 \vee (\ell + 2)K_2$. Then the order of G is $2\ell + 5$ and $\operatorname{diam}_D(G) = 4$. One can also verify that $t(G) = 3\ell + 4 = 2(2\ell + 5) - 6 - \ell$. Similarly, it is not the case that $t(G) = n$ if and only if $\operatorname{diam}_D(G) = n - 2$. Instead, we have the following.

Theorem 3.15. *Let G be a connected graph of order n. If $t(G) = n$, then*

$$2 \leq \lfloor 2n/3 \rfloor \leq \operatorname{diam}_D(G) \leq n - 2.$$

Proof. If $t(G) = n$, then $n \geq 4$ since G is not traceable. The upper bound is an immediate consequence of Theorem 3.5. For the lower bound, let $W = (w_0, w_1, \ldots, w_n)$ be a traceable walk in G. We may assume that $w_i = w_j$ for some i, j with $1 \leq i < j \leq n - 1$ and $j - i \geq 2$. Therefore, G contains three paths $(w_0, w_1, \ldots, w_{j-1})$, $(w_0, w_1, \ldots, w_i, w_{j+1}, w_{j+2}, \ldots, w_n)$, and $(w_{i+1}, w_{i+2}, \ldots, w_n)$ whose lengths equal $j - 1, n + i - j$, and $n - i - 1$, respectively. Hence, the detour diameter of G is at least $\max\{j - 1, n + i - j, n - i - 1\}$. Assume, to the contrary, that $\operatorname{diam}_D(G) < \lfloor 2n/3 \rfloor$. Then

$$2(n - 1) = (j - 1) + (n + i - j) + (n - i - 1) \leq 3(\lfloor 2n/3 \rfloor - 1) \leq 2n - 3,$$

which is impossible. Thus, G contains a path whose length is at least $\lfloor 2n/3 \rfloor$. ☐

The bounds in Theorem 3.15 are sharp. In fact, by adjusting the values of i and j in the proof, we see that for every pair k, n of integers with $n \geq 4$ and $\lfloor 2n/3 \rfloor \leq k \leq n - 2$, there is a connected graph G with $\operatorname{diam}_D(G) = k$ and $|V(G)| = t(G) = n$. For example, if G is a connected graph with $|V(G)| = t(G) = 12$, then $8 \leq \operatorname{diam}_D(G) \leq 10$ by Theorem 3.15. For $8 \leq i \leq 10$, the graph G_i in Fig. 3.2 has order 12, traceable number 12, and $\operatorname{diam}_D(G_i) = i$.

3.3 The Traceable and Hamiltonian Numbers of a Graph

We have already observed that $1 \leq h(G) - t(G) \leq \operatorname{diam}(G)$ for every nontrivial connected graph G. This suggests investigating the value of $h(G) - t(G)$, which is the topic of this section. First, we look at conditions for a graph G under which $h(G) - t(G)$ attains either the lower or upper bound above.

Fig. 3.2 Graphs G_8, G_9, and G_{10}

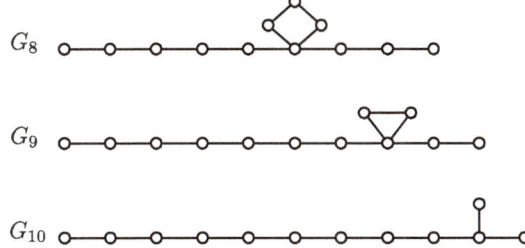

Theorem 3.16 ([53]). *Let G be a nontrivial connected graph. Then $h(G) - t(G) = 1$ if and only if G is either K_2 or Hamiltonian.*

Proof. First, $h(K_2) = t(K_2) + 1 = 2$. If G is a Hamiltonian graph with $n = |V(G)| \geq 3$, then certainly $h(G) = n$ and $t(G) = n - 1$; so $h(G) - t(G) = 1$. To verify the converse, suppose that G is a graph that is not Hamiltonian. Let $W = (w_1, w_2, \ldots, w_{h(G)}, w_1)$ be a Hamiltonian walk in G. Since W contains more than n vertices, we may assume, without loss of generality, that $w_1 = w_i$ for some i with $3 \leq i \leq h(G) - 1$. Then $W' = (w_2, w_3, \ldots, w_{h(G)})$ is a spanning walk in G. Thus, $t(G) \leq L(W') = L(W) - 2 = h(G) - 2$, that is, $h(G) - t(G) \geq 2$. □

Theorem 3.17. *Let G be a nontrivial connected graph of order n. Then $h(G) - t(G) = \text{diam}(G)$ if and only if there exists a cyclic ordering s : $v_1, v_2, \ldots, v_n, v_{n+1} = v_1$ of $V(G)$ such that $d(s) = h(G)$ and $\max\{d(v_i, v_{i+1}) : 1 \leq i \leq n\} = \text{diam}(G)$.*

Proof. Suppose first that $h(G) - t(G) = \text{diam}(G)$. Let $s_\ell : v_1, v_2, \ldots, v_n$ be a linear ordering of $V(G)$ with $d(s_\ell) = t(G)$. Then considering the cyclic ordering s_c : $v_1, v_2, \ldots, v_n, v_1$, we have $t(G) + \text{diam}(G) = h(G) \leq d(s_c) = d(s_\ell) + d(v_1, v_n) \leq t(G) + \text{diam}(G)$, that is, $d(s_c) = h(G)$ and $d(v_1, v_n) = \text{diam}(G)$.

For the converse, suppose that s_c : $v_1, v_2, \ldots, v_n, v_{n+1} = v_1$ is a cyclic ordering of $V(G)$ with $d(s_c) = h(G)$ and, without loss of generality, $d(v_1, v_n) = \text{diam}(G)$. Then for the linear ordering s_ℓ : $v_1, v_2, \ldots, v_n$ of $V(G)$, we have $t(G) \leq d(s_\ell) = d(s_c) - d(v_1, v_n) = h(G) - \text{diam}(G)$. Since $h(G) - t(G) \leq \text{diam}(G)$ by Theorem 3.3, it follows that $h(G) - t(G) = \text{diam}(G)$. □

Theorem 3.18. *Let G be a nontrivial connected graph. If $h(G) - t(G) = \text{diam}(G)$, then the initial and terminal vertices of every traceable walk in G are antipodal vertices.*

Proof. Let $W = (w_0, w_1, \ldots, w_{t(G)})$ be a traceable walk in G. Let $v_i = w_{i-1}$ for $i = 1, 2$. For $3 \leq i \leq n = |V(G)|$, let $v_i = w_{j_i}$, where j_i is the smallest positive integer such that $w_{j_i} \notin \{v_1, v_2, \ldots, v_{i-1}\}$. Then s_ℓ : $v_1, v_2, \ldots, v_n$ is a linear ordering of $V(G)$ with $d(s_\ell) = t(G)$. By considering the cyclic ordering s_c : $v_1, v_2, \ldots, v_n, v_1$ of $V(G)$, we obtain

$$t(G) + \operatorname{diam}(G) = h(G) \le d(s_c) = d(s_\ell) + d(v_1, v_n) \le t(G) + \operatorname{diam}(G),$$

that is, $d(v_1, v_n) = \operatorname{diam}(G)$. Hence, the desired result follows since $v_1 = w_0$ and $v_n = w_{t(G)}$.   □

If $\operatorname{diam}(G) \le 3$, then the converse of Theorem 3.18 holds, which we verify next.

Theorem 3.19. *Let G be a nontrivial connected graph and $\operatorname{diam}(G) \le 3$. If $d(u, v) = \operatorname{diam}(G)$ for every traceable $u - v$ walk in G, then $h(G) - t(G) = \operatorname{diam}(G)$.*

Proof. Suppose that $1 \le h(G) - t(G) < \operatorname{diam}(G)$. If $h(G) - t(G) = 1$, that is, if G is Hamiltonian, then there exists a Hamiltonian $u - v$ path such that $d(u, v) = 1 < \operatorname{diam}(G)$. Hence, assume that $h(G) - t(G) = 2$ and $\operatorname{diam}(G) = 3$. Let $W = (w_1, w_2, \ldots, w_{h(G)}, w_1)$ be a Hamiltonian walk in G. Since G is not Hamiltonian, we may assume that $w_1 = w_i$ for some i with $3 \le i \le h(G) - 1$. Then $W' = (w_2, w_3, \ldots, w_{h(G)})$ is a spanning walk in G whose length equals $h(G) - 2 = t(G)$. Therefore, W' is a traceable walk. Furthermore, $d(w_2, w_{h(G)}) \le 2 < \operatorname{diam}(G)$.   □

While every Hamiltonian graph G satisfies $h(G) - t(G) = 1$, there are also numerous graphs G for which $h(G) - t(G) = \operatorname{diam}(G)$ (trees, complete k-partite graphs that are not Hamiltonian, for example). Furthermore, for each pair k, d of positive integers with $k \le d$, there is a graph G with $h(G) - t(G) = k$ and $\operatorname{diam}(G) = d$. In order to see this, we first present a lemma.

Theorem 2.34 states that the Hamiltonian number of a connected graph G equals the sum of the Hamiltonian numbers of the blocks in G. A connected graph whose order and size are equal is a graph that can be obtained from a tree by adding an edge. Such a graph therefore contains exactly one cycle and is called a unicyclic graph.

Lemma 3.1. *Let G be a unicyclic graph of order n in which the length of the unique cycle is ℓ. Then $h(G) = 2n - \ell$.*

Proof. The result clearly holds if $\ell = n$ and so assume that G itself is not a cycle. Let C be the cycle in G. Then the $n - \ell$ edges not belonging to C are bridges. Therefore, G contains $n - \ell + 1$ blocks, one of which is C and each of the remaining is a copy of P_2. It then follows that $h(G) = h(C) + (n - \ell)h(P_2) = \ell + 2(n - \ell) = 2n - \ell$ by Theorem 2.34.   □

Theorem 3.20 ([53]). *For each pair k, d of integers with $1 \le k \le d$, there exists a connected graph G having diameter d such that $h(G) - t(G) = k$.*

Proof. For $k = d$, let G be a tree having diameter d. Then $h(G) = 2(n - 1)$ while $t(G) = 2(n - 1) - d$, where $n = |V(G)|$, and the result is immediate. Thus, assume that $k < d$. For $k = 1$, a cycle of order $2d$ has the desired property. For $k \ge 2$, let G be the unicyclic graph obtained from a path $(v_1, v_2, \ldots, v_{2d-k+1})$ by joining v_1 and $v_{2(d-k+1)}$. Then $|V(G)| = 2d - k + 1$ and $\mathrm{diam}(G) = d$. We also have $h(G) = 2(2d - k + 1) - 2(d - k + 1) = 2d$ by Lemma 3.1. Since G is traceable, $t(G) = 2d - k$. Thus, $h(G) - t(G) = k$.   □

Since $h(G) \leq t(G) + \mathrm{diam}(G)$ for every nontrivial connected graph G and, trivially, $t(G) \geq \mathrm{diam}(G)$, it follows that

$$t(G) + 1 \leq h(G) \leq 2t(G). \tag{3.1}$$

The lower bound is sharp by Theorem 3.16. For the upper bound, observe that $h(G) = 2t(G)$ if and only if $t(G) = \mathrm{diam}(G)$ if and only if G is a path.

By (3.1), if G is a connected graph whose traceable and Hamiltonian numbers are a and b, respectively, then $a < b \leq 2a$. In fact, every pair a, b of positive integers satisfying $a < b \leq 2a$ is realizable as the traceable number and Hamiltonian number, respectively, of some connected graph.

Theorem 3.21 ([53]). *For each pair a, b of positive integers with $a < b \leq 2a$, there exists a connected graph G with $t(G) = a$ and $h(G) = b$.*

We have also seen that if G is a connected graph of order $n \geq 2$, then $n \leq h(G) \leq 2(n-1)$. Hence, if $t(G) = a$ and $h(G) = b$, then

$$1 \leq n - 1 \leq a < b \leq 2(n-1). \tag{3.2}$$

The next result determines all triples (a, b, n) of integers satisfying (3.2) that can be realized as the traceable number, Hamiltonian number, and order, respectively, of some connected graph.

Theorem 3.22 ([53]). *For each triple (a, b, n) of integers with $1 \leq n - 1 \leq a < b \leq 2(n-1)$, there is a connected graph G of order n such that $t(G) = a$ and $h(G) = b$ if and only if either (i) $b = n = a + 1$ or (ii) $b \geq a + 2$.*

Proof. Let G be a connected graph of order n with $t(G) = a$ and $h(G) = b$. Then $a < b$. In particular, if $b = a + 1$, then G is Hamiltonian by Theorem 3.16 and so $b = n$.

For the converse, assume that a, b, n are integers with $1 \leq n - 1 \leq a < b \leq 2(n-1)$ such that $b = n = a + 1$ or $b \geq a + 2$. If $b = n = a + 1$, then every Hamiltonian graph of order n has the desired property. Also, if $b = 2(n-1)$, then a tree of order n and diameter $2(n-1) - a$ has the desired property. Thus, assume that $a + 2 \leq b \leq 2n - 3$. We consider two cases.

Case 1. $a = n - 1$. Then let G_1 be the graph obtained from a path $P = (v_1, v_2, \ldots,$ $v_{2n-a-1})$ by joining v_1 and v_{2n-b}. Then G_1 is a unicyclic traceable graph of order $2n - a - 1 = n$ and so $t(G_1) = n - 1 = a$. Also, $h(G_1) = 2n - (2n - b) = b$ by Lemma 3.1.

Case 2. $a \geq n$. Then let G_2 be the graph obtained from the graph G_1 of order $2n - a - 1$ produced in Case 1 by (i) joining v_i and v_j for all i, j with $1 \leq i < j \leq 2n - b$ and (ii) adding $a - n + 1$ new vertices and joining each of them to v_{2n-b}. Then $|V(G_2)| = (2n - a - 1) + (a - n + 1) = n$. Since G_2 consists of $b - n + 1$ blocks, namely one copy of K_{2n-b} and $b - n$ copies of P_2, Theorem 2.34 implies that $h(G_2) = (2n - b) + 2(b - n) = b$. Note also that $\operatorname{diam}(G_2) = b - a$ and $\operatorname{diam}_D(G_2) = 2n - a - 2$. Since $h(G_2) - \operatorname{diam}(G_2) \leq t(G_2) \leq 2(n - 1) - \operatorname{diam}_D(G_2)$ by Theorems 3.3 and 3.5, it follows that $t(G_2) = a$.   □

3.4   The Upper Traceable Number of a Graph

For a nontrivial connected graph G, recall that the *upper Hamiltonian number* $h^+(G)$ is defined by $h^+(G) = \max\{d(s)\}$, where the maximum is taken over all cyclic orderings s of $V(G)$. As expected, the *upper traceable number* is defined by

$$t^+(G) = \max\{d(s)\},$$

where the maximum is taken over all linear orderings s of $V(G)$. Consequently, $t(G) \leq t^+(G)$ for every G.

We have seen that the only graphs whose Hamiltonian and upper Hamiltonian numbers are equal are complete graphs and stars. For traceable and upper traceable numbers, the two numbers coincide only when the graph is complete. Clearly $t(K_n) = t^+(K_n) = n - 1$. On the other hand, if G is a connected graph of order $n \geq 3$ that is not complete, then G contains a path (x, y, z) where $xz \notin E(G)$. Let s_1 and s_2 be linear orderings of $V(G)$ whose first three terms are x, y, z and y, x, z, respectively, and the remaining $n - 3$ terms (when $n \geq 4$) are exactly the same. Then $d(s_2') = d(s_1') + 1$ and so $t(G) \leq d(s_1') < d(s_2') \leq t^+(G)$.

Theorem 3.23. *Let G be a nontrivial connected graph. Then $t^+(G) = t(G)$ if and only if G is complete.*

Those graphs G for which $h^+(G) - h(G) = 1$ have been determined in Theorem 2.47. Recall that G is a graph satisfying $h^+(G) - h(G) = 1$ if and only if $G = K_1 \vee H$, where $|V(H)| \geq 3$ and $H \in \{H' = K_{1,\ldots,1,2}, H'' = K_{1,n-2}, \overline{H'}, \overline{H''}\}$. Here, let $n = |V(G)| \geq 4$ and observe that

$$t(G) = t^+(G) - 2 = h(G) - 2 = h^+(G) - 3$$

$$= \begin{cases} n - 1 & \text{if } G = K_1 \vee \overline{H''} \\ 2n - 6 & \text{if } G = K_1 \vee H'' = K_{1,1,n-2} \\ 2n - 5 & \text{if } G = K_1 \vee \overline{H'}. \end{cases}$$

There is a proof similar to that given for Theorem 2.47. Note that $K_1 \vee H' = K_{1,\dots,1,2}$.

Theorem 3.24. *Let G be a connected graph of order n. Then $t^+(G) - t(G) = 1$ if and only if $n \geq 3$ and $G \in \{K_{1,\dots,1,2}, K_{1,n-1}\}$.*

Observe that

$$t(G) = t^+(G) - 1 = h^+(G) - 2 = \begin{cases} n - 1 & \text{if } G = K_{1,\dots,1,2} \\ 2n - 4 & \text{if } G = K_{1,n-1}. \end{cases}$$

Upper Traceable and Upper Hamiltonian Numbers of a Graph

In a nontrivial connected graph G, the distance between any two vertices is at most the diameter of G. Thus, $h^+(G) \leq n \operatorname{diam}(G)$ and $t^+(G) \leq (n - 1) \operatorname{diam}(G)$, where n is the order of G. Furthermore, one can quickly verify that equalities hold in both cases for odd cycles, for example.

A vertex in a connected graph G is a *central vertex* in G if its eccentricity equals the radius $\operatorname{rad}(G)$ of G. If every vertex in G is a central vertex, then $\operatorname{rad}(G) = \operatorname{diam}(G)$ and G is said to be *self-centered*. Suppose that G is a connected graph of order n containing k central vertices ($k \geq 1$). Then for every cyclic ordering s_c of $V(G)$, at least $k + 1$ summands in $d(s_c)$ are at most $\operatorname{rad}(G)$ and the remaining summands are at most $\operatorname{diam}(G)$. Thus, $d(s_c) \leq (k+1) \operatorname{rad}(G) + (n-k-1) \operatorname{diam}(G)$. Similarly, $d(s_\ell) \leq k \operatorname{rad}(G) + (n - k - 1) \operatorname{diam}(G)$ for every linear ordering s_ℓ of $V(G)$.

Theorem 3.25. *If G is a nontrivial connected graph of order n containing k central vertices, then*

$$h^+(G) \leq 2 \operatorname{rad}(G) + (n - 2) \operatorname{diam}(G) \leq (k + 1) \operatorname{rad}(G) + (n - k - 1) \operatorname{diam}(G);$$

$$t^+(G) \leq \operatorname{rad}(G) + (n - 2) \operatorname{diam}(G) \leq k \operatorname{rad}(G) + (n - k - 1) \operatorname{diam}(G).$$

By Theorem 3.25, if $h^+(G) = n \operatorname{diam}(G)$ or $t^+(G) = (n - 1) \operatorname{diam}(G)$, then G is self-centered. Note that the converse is not true (consider even cycles, for example). Also, if T is a nontrivial tree of order n, then either (i) T contains exactly one central vertex and $\operatorname{diam}(G) = 2 \operatorname{rad}(G)$ or (ii) T contains exactly two central vertices and $\text{diam}(G) = 2\,\text{rad}(G) - 1$. Thus, $h^+(T) \leq (n-1)\,\text{diam}(T)$ and $t^+(T) \leq (n-3/2)\,\text{diam}(T)$.

When we consider powers of a given connected graph G, the k-th power G^k is complete if and only if $k \geq d = \text{diam}(G)$. In particular, G^{d-1} is the graph such that $uv \in E(G^{d-1})$ if and only if $d(u,v) < d$, which suggests the following.

Theorem 3.26. *Let G be a connected graph of order n and $\text{diam}(G) = d \geq 2$. Then*
(a) $h^+(G) = nd$ *if and only if $\overline{G^{d-1}}$ is Hamiltonian and*
(b) $t^+(G) = (n-1)d$ *if and only if $\overline{G^{d-1}}$ is traceable.*

Thus, if $h^+(G) = n\,\text{diam}(G)$, then $t^+(G) = (n-1)\,\text{diam}(G)$. The converse holds only for $n \leq 4$. If G is a connected graph of order $n \leq 4$, then G is self-centered if and only if G is complete or $G = C_4$. Hence, either $h^+(G) = n\,\text{diam}(G)$ or $t^+(G) < (n-1)\,\text{diam}(G)$. For $n \geq 5$, construct a graph G of order n from $H = K_{\lfloor(n-1)/2\rfloor,\lceil(n-1)/2\rceil}$ by adding a new vertex x and joining x to two vertices of H belonging to different partite sets. Then $\text{diam}(G) = 2$ and $\overline{G}$ is a traceable graph in which x is a cut-vertex. Thus, $t^+(G) = 2(n-1)$ while $h^+(G) = 2n-1$.

While odd cycles are examples of Hamiltonian graphs G of order n for which $h^+(G) = n\,\text{diam}(G) = t^+(G) + \text{diam}(G)$, there are numerous non-Hamiltonian (and even non-traceable) graphs having this property. For integers $d \geq 2$ and $\ell \geq 3$, let G be a subdivision of $K_{2,\ell}$ such that deleting the two vertices of degree ℓ from G results in $P_d + (\ell-1)P_{d-1}$, the union of ℓ paths. In this case, the order of G equals $\ell(d-1) + 3$ and $\text{diam}(G) = d$. Note that every cycle in G is isomorphic to either C_{2d} or C_{2d+1}. As an example, let us consider the situation where $\ell = 3$. Then G can be constructed from a cycle $C = (v_1, v_2, \ldots, v_{2d+1}, v_1)$ and a path $P = (x_1, x_2, \ldots, x_{d-1})$ by adding the edges $x_1 x_{2d+1}$ and $x_{d-1}v_d$. Let

$$E_1 = \{v_1 v_{d+1}, v_{d-1}v_{2d}, v_d v_{2d}, v_d v_{2d+1}, v_{d+1}v_{2d+1}\}$$

$$E_2 = \{v_i x_{d-i} : 1 \leq i \leq d-1\}$$

$$E_3 = \{v_{\lfloor 3d/2\rfloor+1}x_{\lfloor d/2\rfloor}, v_{\lfloor 3d/2\rfloor+1}x_{\lfloor d/2\rfloor+1}\} \text{ (if } d \geq 3)$$

$$E_4 = \{v_i v_{i-d}, v_i x_{i-d-1} : d+2 \leq i \leq \lfloor 3d/2\rfloor\} \text{ (if } d \geq 4)$$

$$E_5 = \{v_i v_{i-d-1}, v_i x_{i-d} : \lfloor 3d/2\rfloor + 2 \leq i \leq 2d-1\} \text{ (if } d \geq 5).$$

Then the subgraph induced by $\cup_{i=1}^5 E_i$ is a Hamiltonian cycle in $\overline{G^{d-1}}$. Hence, G is a non-Hamiltonian (yet traceable) graph such that $h^+(G) = t^+(G) + \text{diam}(G)$. By investigating the graphs constructed in this manner with $\ell \geq 4$, we see that there are non-traceable graphs G for which $h^+(G) = t^+(G) + \text{diam}(G)$ as well.

By Theorem 3.26, if G is bipartite and either $h^+(G) = n\,\text{diam}(G)$ or $t^+(G) = (n-1)\,\text{diam}(G)$, where $n = |V(G)|$, then $\text{diam}(G)$ cannot be even.

Theorem 3.27. *Let G be a nontrivial connected bipartite graph of order n.*
(a) *If $h^+(G) = n \operatorname{diam}(G)$, then $\operatorname{rad}(G) = \operatorname{diam}(G)$ is odd, n is even, and $G \subseteq K_{n/2,n/2}$.*
(b) *If $t^+(G) = (n-1)\operatorname{diam}(G)$, then $\operatorname{rad}(G) = \operatorname{diam}(G)$ is odd and $G \subseteq K_{\lfloor n/2 \rfloor, \lceil n/2 \rceil}$.*

While the difference between the Hamiltonian number and traceable number of a graph is positive and at most the diameter of the graph, there are similar bounds for the upper Hamiltonian number and upper traceable number.

Theorem 3.28. *For a nontrivial connected graph G,*

$$1 \le h^+(G) - t^+(G) \le \operatorname{rad}(G).$$

Proof. The lower bound is obvious. To verify the upper bound, let $s_c : v_1, v_2, \ldots, v_n, v_1$ be a cyclic ordering of $V(G)$, where $n = |V(G)|$, with $d(s_c) = h^+(G)$. Without loss of generality, assume that v_1 is a central vertex in G. Then observe that $t^+(G) \ge d(s_\ell) = h^+(G) - d(v_1, v_n)$, where $s_\ell : v_1, v_2, \ldots, v_n$ is the linear ordering of $V(G)$ obtained from s_c. Since $d(v_1, v_n) \le \operatorname{rad}(G)$, we have $h^+(G) - t^+(G) \le \operatorname{rad}(G)$. □

Therefore, if G is a graph with $\operatorname{rad}(G) = 1$, that is, if G contains a spanning star, then $h^+(G) - t^+(G) = 1$. Of course, these are not the only graphs having this property; we'll soon see that every nontrivial tree T satisfies $h^+(T) - t^+(T) = 1$.

The Upper Traceable Numbers of Trees

For each edge e in a tree T, the *component number* $\operatorname{cn}(e)$ of e is defined in [23] as the minimum order of a component of $T - e$. Let us write $\operatorname{cn}(T) = \sum_{e \in E(T)} \operatorname{cn}(e)$. Although the exact value of $h^+(T)$ is already known in terms of the *median* $\operatorname{med}(T)$ of T (Theorem 2.53), the following upper bound for the upper Hamiltonian numbers of trees is also useful.

Theorem 3.29 ([23]). *If T is a nontrivial tree, then $h^+(T) \le 2\operatorname{cn}(T)$.*

Proof. Let $s : v_1, v_2, \ldots, v_n, v_{n+1} = v_1$ be an arbitrary cyclic ordering of $V(T)$, where $n = |V(T)|$. For each i ($1 \le i \le n$), let Q_i be the $v_i - v_{i+1}$ path in T. Thus, the n paths $Q_1, Q_2, \ldots, Q_n$ traversed in this order results in a closed spanning walk W whose length equals $d(s)$. If $e \in E(T)$, then for each path Q_i, the edge e occurs at most once and $e \in E(Q_i)$ if and only if v_i and v_{i+1} belong to different components in $T - e$. Thus, each edge e in T occurs in W at most $2\operatorname{cn}(e)$ times. Therefore, $d(s) \le 2\operatorname{cn}(T)$. □

Fig. 3.3 A step in the proof
of Theorem 3.30

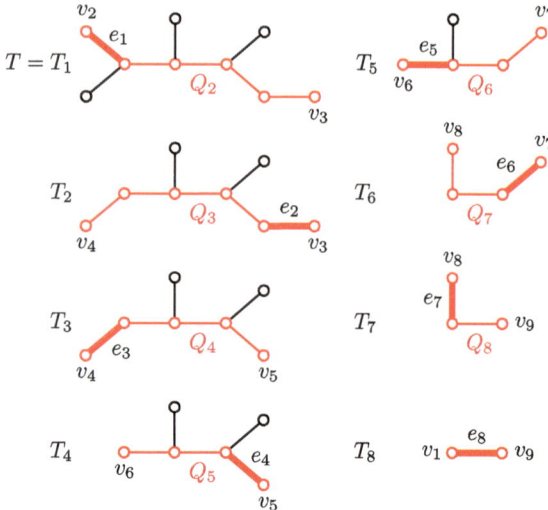

We now present a formula for the upper traceable number of a tree.

Theorem 3.30 ([54]). *If T is a nontrivial tree, then $t^+(T) = 2\operatorname{cn}(T) - 1$.*

Proof. It suffices to find a linear ordering s of $V(T)$ for which $d(s) = 2\operatorname{cn}(T) - 1$
since $t^+(T) < h^+(T) \le 2\operatorname{cn}(T)$. The statement is clearly true if T has order 2.
Hence, we may assume that $n = |V(T)| \ge 3$. Let $T_1 = T$ and suppose that v_2 is
an end-vertex of T_1. Construct a path Q_2 in T_1 with initial vertex v_2 by selecting
every successive edge in Q_2 in such a way that the edge selected at each step has
the maximum component number among all available edges. Suppose that Q_2 is a
$v_2 - v_3$ path. Necessarily, v_3 is another end-vertex of T_1. Let $T_2 = T_1 - v_2$. We then
construct a path Q_3 in T_2 with initial vertex v_3 by selecting each successive edge
in Q_3 so that the edge selected at each step has the maximum component number
among all edges available. The terminal vertex of Q_3 is now called v_4, which is
an end-vertex of T_2, and we next consider $T_3 = T_2 - v_3$. Continuing, we arrive at
the $v_{n-1} - v_n$ path Q_{n-1} in T_{n-2}. The final vertex of T is denoted by v_1, which is
necessarily adjacent to v_{n-1} and v_n. Let T_{n-1} be the tree consisting of v_1 and v_n.
Finally, let Q_1 be the $v_1 - v_2$ path in T. This procedure is illustrated in Fig. 3.3,
where each $v_i - v_{i+1}$ path Q_i for $2 \le i \le n - 1$ is indicated in red.

Thus, $Q_1, Q_2 \subseteq T_1 = T$ while $Q_i \subseteq T_{i-1} = T - \{v_2, v_3, \dots, v_{i-1}\}$ for $3 \le$
$i \le n - 1$. Let e_i be the initial edge in Q_{i+1} for $1 \le i \le n - 2$ and $e_{n-1} = v_1 v_n$.
Then $E(T) = \{e_1, e_2, \dots, e_{n-1}\}$. We will show that $d(s) = 2\operatorname{cn}(T) - 1$ for the
linear ordering $s : v_1, v_2, \dots, v_n$ of $V(T)$ by verifying that each edge $e \in E(T)$ is
traversed either $2\operatorname{cn}(e) - 1$ times or $2\operatorname{cn}(e)$ times by the paths $Q_1, Q_2, \dots, Q_{n-1}$
and the former occurs if and only if $e = e_{n-1}$.

Let $e = e_{i^*}$ be a fixed edge of T. Then $e \in E(T_i)$ if and only if $1 \le i \le i^*$.
In particular, e is a pendant edge in T_{i^*}. For each i with $1 \le i \le \min\{i^*, n - 2\}$,

let S_i' and S_i'' be the components of $T_i - e$ such that $|V(S_i')| \leq |V(S_i'')| + 1$. We claim that if $v_{i+1} \in V(S_i')$, then $v_{i+2} \in V(S_i'')$, that is, the edge e is traversed by the $v_{i+1} - v_{i+2}$ path Q_{i+1}. Let $e = xy$ with $x \in V(S_i')$ and assume, to the contrary, that $v_{i+1}, v_{i+2} \in V(S_i')$. Thus, $x \notin \{v_{i+1}, v_{i+2}\}$.

If $|V(S_i')| \leq |V(S_i'')|$, then the component number of each edge in S_i' is strictly less than $\mathrm{cn}_{T_i}(e)$. Let $(v_{i+1} = u_1, u_2, \ldots, u_k = x)$ and $(v_{i+2} = w_1, w_2, \ldots, w_\ell = x)$ be the $v_{i+j} - x$ paths for $j = 1, 2$, respectively. Obviously, both paths are entirely contained in S_i'. Furthermore, $Q_{i+1} = (v_{i+1} = u_1, u_2, \ldots, u_{k'} = w_{\ell'}, w_{\ell'-1}, \ldots, w_1 = v_{i+2})$ for some integers k' and ℓ' satisfying $2 \leq k' \leq k$ and $2 \leq \ell' \leq \ell$. Then

$$\mathrm{cn}_{T_i}(u_{k'}u_{k'+1}) > \mathrm{cn}_{T_i}(u_{k'-1}u_{k'}) + \mathrm{cn}_{T_i}(w_{\ell'}w_{\ell'-1}) > \mathrm{cn}_{T_i}(w_{\ell'}w_{\ell'-1}).$$

At the same time, however, $k' < k$ and $\mathrm{cn}_{T_i}(u_{k'}u_{k'+1}) \leq \mathrm{cn}_{T_i}(w_{\ell'}w_{\ell'-1})$ by the construction of Q_{i+1}. This is impossible. If $|V(S_i')| = |V(S_i'')| + 1$, then there exists at most one edge in S_i' whose component number equals $\mathrm{cn}_{T_i}(e)$ and the remaining edges in S_i' have component number strictly less than $\mathrm{cn}_{T_i}(e)$. With this, a similar argument given for the case where $|V(S_i')| \leq |V(S_i'')|$ results in another contradiction. Hence, as claimed, $v_{i+2} \in V(S_i'')$ if $v_{i+1} \in V(S_i')$.

Now let T' and T'' be the components of $T - e$, where $|V(T')| \leq |V(T'')|$. Then we may assume that $V(T') = \{v_{j_1}, v_{j_2}, \ldots, v_{j_c}\}$, where $c = \mathrm{cn}(e)$ and $j_1 < j_2 < \cdots < j_c$. Necessarily, the vertex v_{j_c} is incident with e. In each tree T_i containing e (that is, for $1 \leq i \leq i^*$), let T_i' and T_i'' be the components of $T_i - e$ such that $T_i' \subseteq T'$ and $T_i'' \subseteq T''$. Then by the claim given above, it follows that $|V(T_i')| \leq |V(T_i'')|$. Thus, $j_c = i^* + 1$ and no two vertices in T' are consecutive in s. Also, $v_1 \in V(T'')$ and so $j_1 \geq 2$. Hence, if $1 \leq i^* \leq n-2$, then the edge $e = e_{i^*}$ is traversed by the $2c$ distinct paths

$$Q_{j_1-1}, Q_{j_1}, Q_{j_2-1}, Q_{j_2}, \ldots, Q_{j_c-1} = Q_{i^*}, Q_{j_c} = Q_{i^*+1}.$$

Otherwise, $e = e_{n-1}$ and this edge is traversed by the $2c - 1$ distinct paths

$$Q_{j_1-1}, Q_{j_1}, Q_{j_2-1}, Q_{j_2}, \ldots, Q_{j_c-1-1}, Q_{j_c-1}, Q_{j_c-1} = Q_{i^*} = Q_{n-1}.$$

This completes the proof.   □

Theorem 3.31. *If T is a nontrivial tree, then $h^+(T) = t^+(T) + 1 = 2\,\mathrm{cn}(T)$.*

Theorem 3.32. *If T is a nontrivial tree, then $\mathrm{med}(T) = \mathrm{cn}(T)$.*

Recall that upper and lower bounds for the upper Hamiltonian number of a tree were established in terms of its order (Theorem 2.44), that is, if T is a tree of order $n \geq 3$, then $2(n-1) \leq h^+(T) \leq \lfloor n^2/2 \rfloor$. Moreover, $h^+(T) = 2(n-1)$ if and only if T is a star while $h^+(T) = \lfloor n^2/2 \rfloor$ if and only if T is a path. Thus, the following is a consequence of Theorem 3.31.

Theorem 3.33. *Let T be a tree of order $n \geq 3$. Then $2n-3 \leq t^+(T) \leq \lfloor n^2/2 \rfloor - 1$. Furthermore, $t^+(T) = 2n - 3$ if and only if T is a star and $t^+(T) = \lfloor n^2/2 \rfloor - 1$ if and only if T is a path.*

Theorem 3.34. *If T is a nontrivial tree of order n and diameter d, then*

$$2n - 3 + 2\lfloor d/2 \rfloor \lfloor (d-1)/2 \rfloor \leq t^+(T) \leq (n - 3/2)d.$$

Proof. The upper bound has been verified after Theorem 3.25. For the lower bound, let P be a longest path in T. Then observe that

$$\sum_{e \in E(P)} cn_T(e) \geq \begin{cases} 2(1 + 2 + \cdots + d/2) & \text{if } d \text{ is even} \\ 2(1 + 2 + \cdots + (d-1)/2) + (d+1)/2 & \text{if } d \text{ is odd} \end{cases}$$

$$= \lceil d/2 \rceil \lceil (d+1)/2 \rceil.$$

Since $cn(T) \geq n - d - 1 + \sum_{e \in E(P)} cn_T(e)$, the result follows by Theorem 3.30. □

The Traceable and Upper Traceable Numbers of a Graph

We have seen that if G is a nontrivial connected graph of order n, then

$$n - 1 \leq t(G) \leq t^+(G) \leq \lfloor n^2/2 \rfloor - 1$$

and $t(G) = n - 1$ if and only if G contains P_n as a subgraph while $t^+(G) = \lfloor n^2/2 \rfloor - 1$ if and only if G itself is P_n. We say that the ordered pair (a, b) of positive integers is realizable if there exists a graph whose traceable number and upper traceable number are a and b, respectively. Thus, $(a, \lfloor (a+1)^2/2 \rfloor - 1)$ is realizable for every positive integer a. It is then natural to ask which ordered pairs are realizable. A complete answer is established in [36].

Theorem 3.35. *For a pair a, b of positive integers, there exists a graph G such that $(t(G), t^+(G)) = (a, b)$ if and only if either*
  i. *$a \leq b \leq \lfloor a^2/2 \rfloor$ or*
 ii. *$\lfloor a^2/2 \rfloor + 1 \leq b \leq \lfloor (a+1)^2/2 \rfloor - 1$ and b is odd.*

3.5   Traceable Numbers of Vertices in a Graph

Since P_n and C_n $(n \geq 3)$ are, of course, both traceable graphs of order n, it follows that $t(P_n) = t(C_n) = n - 1$. One difference between these two graphs is that, for P_n, there are exactly two linear orderings s of the vertices for which $d(s) = n - 1$, while for C_n, there are $2n$ such linear orderings. More specifically, the initial vertex

Fig. 3.4 A graph G with
$t(G) = 4$ and $t^{\star}(G) = 5$

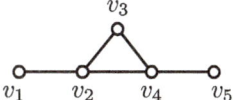

of every Hamiltonian path in P_n must be one of the two end-vertices. In contrast, every vertex in C_n is the initial vertex of some Hamiltonian path.

Let G be a nontrivial connected graph of order n. For a vertex v of G, the *traceable number $t(v)$* of v is defined by

$$t(v) = \min\{d(s)\},$$

where the minimum is taken over all linear orderings s of $V(G)$ *whose initial term is v.* Therefore, $t(v) \geq n - 1$ for every $v \in V(G)$ and $t(v) = n - 1$ if and only if G contains a Hamiltonian path with initial vertex v. Observe also that the traceable number $t(G)$ of G can be alternatively defined by

$$t(G) = \min\{t(v) : v \in V(G)\}.$$

As in Theorems 2.26 and 3.2, we have the following.

Theorem 3.36 ([53]). *Let G be a nontrivial connected graph and $v \in V(G)$. Then $t(v)$ equals the length of a shortest spanning walk in G whose initial vertex is v.*

For example, in the traceable graph G shown in Fig. 3.4, every Hamiltonian path starts at one of the two end-vertices (and ends at the other end-vertex). Hence, $t(v_1) = t(v_5) = t(G) = 4$ while $t(v_i) \geq 5$ for $2 \leq i \leq 4$. In fact, $t(v_i) = 5$ for $2 \leq i \leq 4$ since $t(v_i) \leq d(s_i) = 5$ where

$$s_2 : v_2, v_1, v_3, v_4, v_5$$

$$s_3 : v_3, v_1, v_2, v_4, v_5$$

$$s_4 : v_4, v_5, v_3, v_2, v_1.$$

For a nontrivial connected graph G, the *maximum vertex traceable number $t^{\star}(G)$* of G is defined as

$$t^{\star}(G) = \max\{t(v) : v \in V(G)\}.$$

Thus, $t(G) = 4$ and $t^{\star}(G) = 5$ for the graph G in Fig. 3.4. Obviously, $t(G) \leq t^{\star}(G)$ for every connected graph G. The maximum vertex traceable numbers of graphs will be discussed further in Sect. 3.6.

For two vertices u and v in a connected graph G, we have $0 \leq |t(u) - t(v)| \leq t^{\star}(G) - t(G)$ as both $t(u)$ and $t(v)$ are between $t(G)$ and $t^{\star}(G)$. In fact, we have the following.

Theorem 3.37 ([53]). *Let G be a nontrivial connected graph and $u, v \in V(G)$. Then $|t(u) - t(v)| \leq d(u, v)$.*

Proof. It suffices to show that $|t(u) - t(v)| \leq 1$ when $uv \in E(G)$. Let $n = |V(G)|$ and consider a linear ordering $s : v = v_1, v_2, \ldots, v_n$ such that $d(s) = t(v)$. Therefore, $u = v_i$ for some i with $2 \leq i \leq n$. If $u = v_n$, then consider the linear ordering $s' : u = v_n, v_{n-1}, \ldots, v_1$. Clearly, $t(u) \leq d(s') = d(s) = t(v)$. Otherwise, let $s'' : u = v_i, v_{i-1}, \ldots, v_1 = v, v_{i+1}, v_{i+2}, \ldots, v_n$ and observe that

$$t(u) \leq d(s'') = d(s) + d(v, v_{i+1}) - d(u, v_{i+1})$$
$$\leq t(v) + (d(v, u) + d(u, v_{i+1})) - d(u, v_{i+1})$$
$$= t(v) + 1.$$

Thus, $t(u) - t(v) \leq 1$ in each case. One can similarly show that $t(v) - t(u) \leq 1$. □

For a vertex v in a connected graph G, recall that the eccentricity $e(v)$ is the greatest distance between v and a vertex in G.

Theorem 3.38 ([53]). *If v is a vertex in a tree of order $n \geq 2$, then*

$$t(v) = 2(n - 1) - e(v).$$

Proof. We first show that $t(v) \geq 2(n - 1) - e(v)$, where v is a vertex in a tree T of order $n \geq 2$. Let $s : v = v_1, v_2, \ldots, v_n$ be a linear ordering of $V(T)$ such that $d(s) = t(v)$. Then for the cyclic ordering $s' : v_1, v_2, \ldots, v_n, v_1$ of $V(T)$, we have

$$2(n - 1) = h(T) \leq d(s') = d(s) + d(v_1, v_n) \leq t(v) + e(v)$$

and so $t(v) \geq 2(n - 1) - e(v)$.

In order to show that $t(v) \leq 2(n - 1) - e(v)$ for each vertex v in a nontrivial tree of order n, we proceed by induction on n. The result clearly holds for $n = 2$. For an integer $n \geq 3$, assume that $t(u) \leq 2n - 4 - e(u)$ for every tree T' of order $n - 1$ and every vertex u in T'. Let T be a tree of order n and $v \in V(T)$. If $T = P_n$ and v is an end-vertex, then certainly $t(v) = e(v) = n - 1$ and so $t(v) = 2(n - 1) - e(v)$. Therefore, let us assume that this is not the case. Let P be a longest path in T with initial vertex v, say P is a $v - u$ path. Therefore, $e(v) = d(v, u)$. Let w be an end-vertex that does not belong to P. Therefore, $T' = T - w$ is a tree of order $n - 1$ containing v and $e_{T'}(v) = e_T(v)$. By the induction hypothesis,

$$t_{T'}(v) \leq 2n - 4 - e_{T'}(v) = 2n - 4 - e_T(v).$$

Let x be the vertex that is adjacent to w in T. If $s' : v = v_1, v_2, \ldots, v_{n-1}$ is a linear ordering of $V(T')$ such that $d(s') = t_{T'}(v)$, then $x = v_i$ for some i with $1 \leq i \leq n - 1$. Now let s be the linear ordering of $V(T)$ obtained from s' by inserting w immediately after v_i. Then one can verify that $d(s) = d(s') + 1$ if $x = v_{n-1}$ and $d(s) \le d(s') + 2$ otherwise. Hence, $t_T(v) \le d(s) \le d(s') + 2 = t_{T'}(v) + 2 \le 2(n-1) - e_T(v)$. $\qquad\square$

By Theorem 3.38, if T is a tree of order $n \ge 2$, then

$$t(T) = \min\{t(v) : v \in V(T)\}$$
$$= \min\{2(n-1) - e(v) : v \in V(T)\}$$
$$= 2(n-1) - \max\{e(v) : v \in V(T)\}$$
$$= 2(n-1) - \mathrm{diam}(T),$$

which provides us with an alternative proof of Theorem 3.9.

3.6   The Maximum Vertex Traceable Number of a Graph

As defined in Sect. 3.5, the *maximum vertex traceable number* $t^\star(G)$ of G is defined as

$$t^\star(G) = \max\{t(v) : v \in V(G)\}.$$

Clearly $t(G) \le t^\star(G) \le t^+(G)$. The following is a consequence of Theorems 3.23 and 3.24.

Theorem 3.39. *If G is a nontrivial connected graph of order n, then*
(a) $t(G) = t^\star(G) = t^+(G)$ *if and only if G is complete,*
(b) $t(G) = t^\star(G) = t^+(G) - 1$ *if and only if $n \ge 4$ and $G = K_{1,\dots,1,2}$, and*
(c) $t(G) + 1 = t^\star(G) = t^+(G)$ *if and only if $n \ge 3$ and G is a star.*

Trees form another class of graphs for which the maximum vertex traceable numbers are easy to find. If T is a nontrivial tree of order n, then

$$t^\star(T) = \max\{t(v) : v \in V(T)\}$$
$$= \max\{2(n-1) - e(v) : v \in V(T)\}$$
$$= 2(n-1) - \mathrm{rad}(T).$$

Theorem 3.40. *If T is a nontrivial tree of order n, then*

$$t(T) = 2(n-1) - \mathrm{diam}(T) \ \ and \ \ t^\star(T) = 2(n-1) - \mathrm{rad}(T).$$

If n, r are integers such that $1 \le r \le \lceil (n-1)/2 \rceil$, then there exists a tree T of order n and radius r.

Theorem 3.41. *For each pair n, k of integers with $n \geq 2$ and*

$$\lfloor 3(n-1)/2 \rfloor \leq k \leq 2(n-1) - 1,$$

there exists a tree T of order n with $t^\star(T) = k$.

If G is a nontrivial connected graph of order n and $t^\star(G) = n - 1$, then $t(v) = n - 1$ for every $v \in V(G)$, that is, every vertex is the initial vertex of a Hamiltonian path. Such a graph G is called homogeneously traceable. Homogeneously traceable graphs will be discussed further later.

If H is a connected spanning subgraph of a nontrivial graph G, then $t^\star(G) \leq t^\star(H)$. Thus, if G is a connected graph of order $n \geq 2$ and $t^\star(G) = 2n - 3$, then $K_{1,n-1}$ is the unique spanning tree of G. Therefore, either G itself is a star or $G = K_3$. Since $t^\star(K_3) = 2 < 2 \cdot 3 - 3$, we have the following.

Theorem 3.42. *If G is a nontrivial connected graph of order n, then*

$$n - 1 \leq t(G) \leq t^\star(G) \leq 2n - 3.$$

Also, $t^\star(G) = n - 1$ if and only if G is homogeneously traceable while $t^\star(G) = 2n - 3$ if and only if $G = K_{1,n-1}$.

For a given nontrivial connected graph G, it is clear that $t(G) \leq t^\star(G) \leq t^+(G)$. Also,

$$t^\star(G) = t^+(G) \text{ if } G \text{ is either a complete graph or a star.} \qquad (3.3)$$

Thus, $t^+(G)$ is a sharp upper bound for $t^\star(G)$. However, in most instances, there is an improved upper bound for $t^\star(G)$ in terms of $t^+(G)$ and another graphical parameter, namely $t(G)$.

Theorem 3.43. *For every nontrivial connected graph G,*

$$t^\star(G) \leq (t(G) + t^+(G) + 1)/2.$$

Proof. Since the result is immediate when $t(G) = t^\star(G)$, we assume that $t(G) < t^\star(G)$. It suffices to show that $t^\star(G) - t(G) \leq t^+(G) - t^\star(G) + 1$. Let $s_0 : v_1, v_2, \ldots, v_n$ be a linear ordering of $V(G)$, where $n = |V(G)|$, for which $d(s_0) = t(G)$. Let x be a vertex for which $t(x) = t^\star(G)$. Since $t(G) \neq t^\star(G)$, we may assume that $x = v_i$ for some i, where $2 \leq i \leq n - 1$. Consider the linear ordering $s_1 : x, v_1, v_2, \ldots, v_{i-1}, v_{i+1}, v_{i+2}, \ldots, v_n$ and observe that

$$t^\star(G) = t(x) \leq d(s_0) + d(x, v_1) + d(v_{i-1}, v_{i+1}) - d(v_{i-1}, v_i) - d(v_i, v_{i+1})$$
$$\leq t(G) + e(x),$$

where $e(x)$ is the eccentricity of x. Now let y and z be a neighbor of x and a vertex farthest from x, respectively. For a linear ordering s' of $V(G) - \{x\}$ whose first and last terms are y and z, respectively, let s_2 and s_3 be the linear orderings of $V(G)$ such that s_2 is the ordering whose initial term is x followed by s' and s_3 is the ordering whose terminal term is x preceded by s'. Then $t^\star(G) \leq d(s_2) = d(s') + 1 \leq d(s') + e(x) = d(s_3) \leq t^+(G)$ and so $t^+(G) - t^\star(G) + 1 \geq e(x)$. Hence, $t^\star(G) - t(G) \leq e(x) \leq t^+(G) - t^\star(G) + 1$.   □

By Theorem 3.43, if G is a graph satisfying $t^\star(G) = t^+(G)$, then $t^+(T) \leq t(G) + 1$. Hence, Theorem 3.39 implies that the converse of (3.3) also holds, that is, $t^\star(G) = t^+(G)$ if and only if G is either a complete graph or a star.

Problem 3.1. For which graphs G is $t^\star(G) = (t(G) + t^+(G) + 1)/2$?

We saw in Theorem 3.37 that $|t(u) - t(v)| \leq 1$ when u and v are adjacent. The following is an immediate consequence of this observation.

Theorem 3.44. *Let G be a nontrivial connected graph. If k is an integer such that $t(G) \leq k \leq t^\star(G)$, then there exists a vertex whose traceable number equals k.*

Proof. Since the statement is obvious if $k = t(G)$ or $k = t^\star(G)$, assume that $t(G) < k < t^\star(G)$. Let $u, v \in V(G)$ such that $t(u) = t(G)$ and $t(v) = t^\star(G)$. Consider a $u - v$ geodesic $P = (u = v_0, v_1, \ldots, v_\ell = v)$, where $\ell = d(u, v)$. Suppose that j is the largest integer such that v_j belongs to P and $t(v_j) \leq k$. Thus, $j \leq \ell - 1$. We claim that $t(v_j) = k$. If this is not the case, then $t(v_j) \leq k - 1$ and $t(v_{j+1}) \geq k + 1$. However then, $t(v_{j+1}) - t(v_j) \geq 2$, which contradicts Theorem 3.37.   □

The *traceable vertex spectrum* $\mathcal{T}(G)$ of a nontrivial connected graph G is defined as $\mathcal{T}(G) = \{t(v) : v \in V(G)\}$. Thus, Theorem 3.44 implies that $\mathcal{T}(G) = \{t(G), t(G) + 1, t(G) + 2, \ldots, t^\star(G)\}$. In particular, if $t(G) = t^\star(G)$, then $\mathcal{T}(G)$ is a singleton set and so $t(v)$ is a constant for all $v \in V(G)$. For this reason, a graph G is said to be *traceably singular* if $t(G) = t^\star(G)$. It is therefore natural to ask what properties traceably singular graphs possess. Note that if G is a Hamiltonian graph of order $n \geq 3$, then every vertex is the initial vertex of a Hamiltonian path, that is, $\mathcal{T}(G) = \{n - 1\}$. Therefore, every Hamiltonian graph is certainly traceably singular. There are also non-Hamiltonian graphs with this property. For example, although the Petersen graph P is not Hamiltonian, every vertex is the initial vertex of a Hamiltonian path and so $\mathcal{T}(P) = \{9\}$. In fact, deleting any edge or vertex from P results in another traceably singular graph. On the other hand, deleting any two edges from P results in a graph that is not traceably singular.

The Petersen graph is an example of a hypohamiltonian graph. A graph G is *hypohamiltonian* if G itself is not Hamiltonian but $G - v$ is Hamiltonian for every $v \in V(G)$. Every hypohamiltonian graph is 3-connected since deleting any two

Fig. 3.5 The graph G^*

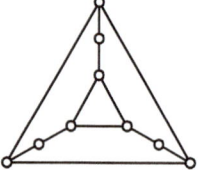

vertices results in a traceable graph. Furthermore, hypohamiltonian graphs form another class of graphs that are not only traceable but also traceably singular.

Theorem 3.45. *If G is a graph that is either Hamiltonian or hypohamiltonian, then G is traceable and traceably singular.*

It is known that the Petersen graph is the smallest hypohamiltonian graph (see [41]). Among the known infinite classes of hypohamiltonian graphs is one found by Lindgren [49]. For each positive integer k, let H_k be the graph obtained from a $(6k+3)$-cycle $C = (v_1, v_2, \ldots, v_{6k+3}, v_1)$ by adding the edges $v_i v_j$, where $1 \leq i \leq 3k+1$ and either (i) $i \equiv 1 \pmod 3$ and $j - i = 3k+1$ or (ii) $i \equiv 2 \pmod 3$ and $j - i = 3k+2$. Thus, H_k is a Hamiltonian graph of order $6k+3$ containing $2k+1$ vertices of degree 2 and $4k+2$ vertices of degree 3. From H_k, the graph G_k is obtained by adding a new vertex x and the edges xv_i if and only if $i \equiv 0 \pmod 3$. Therefore, G_k is a graph of order $6k+4$ such that $\deg v = 3$ if $v \in V(C)$ and $\deg v = 2k+1$ if $v = x$. The graph H_1 is the Petersen graph. By symmetry, it is not too difficult to verify that $G_k - v$ is Hamiltonian for every $v \in V(G_k)$. Furthermore, while H_k is clearly Hamiltonian, if there is a $u - v$ Hamiltonian path in H_k, then at most one of u and v is adjacent to x in G_k, which implies that G_k is not Hamiltonian.

Another example of a traceably singular graph is the graph G^* shown in Fig. 3.5. Note that G^* is neither Hamiltonian nor hypohamiltonian. This is a graph of order 9 and size 12 in which every vertex is the initial vertex of a Hamiltonian path. In fact, there exists a Hamiltonian $u - v$ path if and only if either (i) $\deg u = 2$ and $d(u, v) \geq 2$ or (ii) $\deg u = 3$ and $d(u, v) = \deg v$.

With the aid of G^*, one can construct a traceably singular graph of order n that is neither Hamiltonian nor hypohamiltonian for every integer $n \geq 9$.

Theorem 3.46. *For each integer $n \geq 9$, there exists a traceably singular graph of order n that is neither Hamiltonian nor hypohamiltonian.*

Proof. Since the graph G^* has the desired property, we may assume that $n \geq 10$. Let v be a vertex in G^* whose degree equals 2 and let $N(v) = \{u, w\}$. Consider the graph G_n obtained from $G^* - v$ by adding a copy of K_{n-8} and joining each of these $n - 8$ vertices to both u and w. One can then verify that G_n has the desired properties. □

Traceably Singular Graphs That Are Traceable: Homogeneously Traceable Graphs

The examples of traceably singular graphs we have seen thus far are all traceable. A graph G is *homogeneously traceable* if every vertex is the initial vertex of a Hamiltonian path in G. Thus, a homogeneously traceable graph is traceably singular and $t(G) = t^*(G) = n - 1$, where $n = |V(G)|$. Since Hamiltonian graphs are obviously homogeneously traceable, it is more interesting to restrict our attention to homogeneously traceable graphs that are not Hamiltonian. These graphs have been studied by Skupień [63] and Chartrand, Gould, and Kapoor [22].

Suppose that G is a nontrivial homogeneously traceable graph. Since every vertex is the initial vertex of a Hamiltonian path in G, it follows that $G - v$ is traceable for every $v \in V(G)$. Hence, G is clearly 2-connected. Furthermore, $G - v$ contains at most two end-vertices. In particular, when G is not Hamiltonian, then $G - v$ contains at most one end-vertex. These facts are summarized as follows.

Theorem 3.47 ([22, 63]). *Let G be a nontrivial homogeneously traceable graph. Then G is 2-connected and $k(G - S) \le |S|$ for every nonempty proper subset S of $V(G)$. For every vertex $v \in V(G)$, the graph $G - v$ is traceable and contains at most two end-vertices.*

Theorem 3.48 ([22]). *Let G be a nontrivial homogeneously traceable graph that is not Hamiltonian. Then for every $v \in V(G)$, the graph $G - v$ is traceable and contains at most one end-vertex. Also, the number of vertices of degree 2 is at most $|V(G)|/2$.*

The proof of Ore's Theorem (Theorem 2.2) implies the following.

Lemma 3.2. *Let G be a graph of order $n \ge 3$ and $u, v \in V(G)$. If there exists a Hamiltonian $u - v$ path and $\deg u + \deg v \ge n$, then G is Hamiltonian.*

Theorem 3.49 ([22]). *If G is a non-Hamiltonian homogeneously traceable graph of order n, then (i) $n \in \{1, 2\}$ or (ii) $n \ge 7$ and $3 \le \Delta(G) \le n - 4$.*

Proof. The result is obvious for $n = 1, 2$. Also, if G is a 2-connected graph of order $n \in \{3, 4, 5\}$, then either G is Hamiltonian or $G \in \{K_{2,3}, K_{1,1,3}\}$. Since neither $K_{2,3}$ nor $K_{1,1,3}$ is homogeneously traceable, the result holds for $3 \le n \le 5$ as well. Thus, let us assume that $n \ge 6$. Let v be a vertex whose degree equals $\Delta(G)$ and consider a Hamiltonian path $(v = v_1, v_2, \ldots, v_n)$ whose initial vertex is v. Since $\deg v_n \ge 2$, it follows that G is Hamiltonian or $\Delta(G) \le n - 3$ by Lemma 3.2. Now assume, to the contrary, that G is a non-Hamiltonian homogeneously traceable graph and $\Delta(G) = n - 3$. Since $\deg v = n - 3$, we have $\deg v_n = 2$. Suppose that $v_i v_n \in E(G)$, where $2 \le i \le n - 2$. Thus, there is a Hamiltonian $v - v_{i+1}$ path, implying that $\deg v_{i+1} = 2$. If $i = n - 2$, then v_{n-2} is a cut-vertex, which cannot occur. Also, if $i \leq n - 3$, then $G - v_{n-3}$ contains two end-vertices, which is again impossible. Thus, $\Delta(G) \leq n - 4$. Since $G \neq C_n$, the result now follows. $\qquad\square$

Theorem 3.50 ([22]). *If G is a non-Hamiltonian homogeneously traceable graph of order n and $\Delta(G) = n - 4$, then the subgraph induced by the vertices of degree $n - 4$ is complete.*

Proof. Assume, to the contrary, that there exists a non-Hamiltonian homogeneously traceable graph G of order n containing nonadjacent vertices u, v of degree $n - 4$. Add as many edges as possible between pairs of nonadjacent vertices of G so that the resulting graph H is still not Hamiltonian. Hence, H is a non-Hamiltonian homogeneously traceable graph of order n such that addition of an edge to H produces a Hamiltonian graph. Note that $\deg_H u = \deg_H v = n - 4 = \Delta(H)$. Since $H + uv$ is Hamiltonian, it follows that H contains a Hamiltonian $u - v$ path. Therefore, $2(n - 4) = \deg_H u + \deg_H v \leq n - 1$, that is, $n = 7$. We may assume that $(v_1, v_2, \ldots, v_7)$ is a Hamiltonian path in H and $\deg v_1 = \deg v_7 = 3$ while $v_1 v_7 \notin E(H)$. Let $N(v_1) = \{v_2, v_{i_1}, v_{i_2}\}$, where $3 \leq i_1 < i_2 \leq 6$, and $N(v_7) = \{v_6, v_{j_1}, v_{j_2}\}$, where $2 \leq j_1 < j_2 \leq 5$. Now observe that $\{i_1, i_2\} \cap \{j_1, j_2\} = \emptyset$ since $\Delta(H) = 3$. Also, $\{i_1, i_2\} \cap \{j_1 + 1, j_2 + 1\} = \emptyset$ since H is not Hamiltonian. Therefore, $\{i_1, i_2\} = \{5, 6\}$ and $\{j_1, j_2\} = \{2, 3\}$. However, this still produces a Hamiltonian cycle in H, which is a contradiction. $\qquad\square$

It is now easy to verify that every homogeneously traceable graph of order 7 must be Hamiltonian.

Theorem 3.51 ([22]). *Let G be a graph of order 7. Then G is Hamiltonian if and only if G is homogeneously traceable.*

Proof. If there exists a non-Hamiltonian homogeneously traceable graph G of order 7, then G contains at most three vertices of degree 2 by Theorem 3.50. Then by Theorem 3.49, there are at least four vertices of degree 3. However, this is impossible by Theorem 3.50. $\qquad\square$

Similarly, there is no non-Hamiltonian homogenously traceable graph of order 8. If there were such a graph G, then $2 \leq \deg v \leq 4$ for every $v \in V(G)$. It is stated in [22] that consideration of all possible cases yields the result that no such G exists.

While a graph that is homogeneously traceable must have sufficiently many edges, having too many edges results in a Hamiltonian graph. By Theorem 3.48, if G is a homogeneously traceable non-Hamiltonian graph of order $n \geq 9$, then at most $n/2$ vertices in G are of degree 2 while the remaining vertices are of degree greater than 2.

Theorem 3.52 ([22]). *If G is a homogeneously traceable non-Hamiltonian graph of order n, then either (i) $n = 1, 2$ or (ii) $n \geq 9$ and $|E(G)| \geq 5n/4$.*

It is also shown in [22] that this is a sharp bound. For example, let ℓ be a positive integer and consider the graph $H = C_{2\ell+1} \square P_2$ (the Cartesian product of $C_{2\ell+1}$ and P_2). Let C and C' be the two vertex-disjoint copies of $C_{2\ell+1}$ in H. Then G is obtained from H by replacing each P_2 joining $v \in V(C)$ and $v' \in V(C')$ by P_4 such that $d_G(v, v') = 3$. Observe then that the resulting graph G has order $4(2\ell + 1)$ and size $5(2\ell + 1)$, possessing the desired properties.

Randomly Traceable Graphs

A traceable graph G is said to be *randomly traceable* if a Hamiltonian path always results upon starting at any vertex in G and successively proceeding to any adjacent vertex not yet encountered. In other words, for an arbitrary $u - v$ path P in G, there exists a Hamiltonian path P' with initial vertex u such that $P \subseteq P'$. Thus, a randomly traceable graph is homogeneously traceable but the converse is false. (Consider, for example, the Petersen graph.) If, in addition, $|V(G)| \geq 3$ and the final vertex of each such path is adjacent to the first vertex, then G is called *randomly Hamiltonian*. These graphs were studied by Chartrand and Kronk [16].

While a randomly Hamiltonian graph is clearly randomly traceable, if G is a randomly traceable graph of order at least 3, then G turns out to be randomly Hamiltonian as well.

Theorem 3.53 ([16]). *Let G be a graph of order at least 3. Then G is randomly traceable if and only if G is randomly Hamiltonian.*

Proof. Let G be a randomly traceable graph of order n and suppose that $P = (v_1, v_2, \ldots, v_n)$ is a Hamiltonian path. Consider the subpath $P' = (v_2, v_3, \ldots, v_n)$ of P. Since G is randomly traceable, it must be possible to begin with P' and conclude with a Hamiltonian path. This implies that $v_1 v_n \in E(G)$, that is, P can be extended to a Hamiltonian cycle. $\square$

Thus, given any randomly traceable graph G of order $n \geq 3$, we may assume the presence of a Hamiltonian cycle. More can be said.

Lemma 3.3 ([16]). *Let G be a randomly traceable graph of order $n \geq 3$ with a Hamiltonian cycle $C = (v_1, v_2, \ldots, v_n, v_1)$. For each integer ℓ with $1 \leq \ell \leq n/2$, let $E_\ell = \{v_i v_{i+\ell} : 1 \leq i \leq n\}$, where the subscripts are expressed modulo n. Then either $E_\ell \subseteq E(G)$ or $E_\ell \cap E(G) = \emptyset$.*

Proof. Since $E_1 = E(C) \subseteq E(G)$, we may assume that $\ell \geq 2$. It suffices to show that if $v_i v_j \in E_\ell \cap E(G)$, then $v_{i+1} v_{j+1} \in E(G)$. Also, we may assume, without loss of generality, that $i = 1$. Observe that $(v_2, v_3, \ldots, v_j, v_1, v_n, v_{n-1}, \ldots, v_{j+1})$ is a Hamiltonian path. Since G is randomly Hamiltonian, it follows that $v_2 v_{j+1} \in E(G)$. $\square$

By Lemma 3.3, if G is a randomly traceable graph of order at most 5, then G must be either a cycle or a complete graph. For graphs of order at least 6, the following is also useful.

Lemma 3.4. *Let G be a randomly traceable graph of order $n \geq 6$ with a Hamiltonian cycle $C = (v_1, v_2, \ldots, v_n, v_1)$. If ℓ is an integer with $3 \leq \ell \leq n - 3$, then $v_1 v_\ell \in E(G)$ implies that $v_1 v_{\ell+2} \in E(G)$.*

Proof. Assume that $v_1 v_\ell \in E(G)$. Then by Lemma 3.3, G contains a path $P = (v_{\ell+3}, v_{\ell+4}, \ldots, v_n, v_{\ell-1}, v_{\ell-2}, \ldots, v_2, v_{\ell+1}, v_\ell, v_1)$, which contains every vertex in G except for $v_{\ell+2}$. Since we must be able to extend P to a Hamiltonian path, it follows that $v_1 v_{\ell+2} \in E(G)$. $\square$

Theorem 3.54 ([16]). *A nontrivial graph G is randomly traceable if and only if G is a cycle, a complete graph, or a regular complete bipartite graph.*

Proof. It is straightforward to verify that cycles, complete graphs, and regular complete bipartite graphs are randomly traceable.

Conversely, let G be a randomly traceable graph of order n. For $n = 2, 3$, the result is obvious. Hence, assume that $n \geq 4$ and let $C = (v_1, v_2, \ldots, v_n, v_1)$ be a Hamiltonian cycle in G. If $G = C$, then $G = C_n$. We therefore assume that G contains edges not belonging to C (called chords) and necessarily then cycles containing exactly one chord. Such cycles will be referred to as outer cycles. Consider the smallest p for which G contains an outer p-cycle.

Let p be the smallest positive integer such that $v_1 v_p \in E(G)$. (Thus, $2 \leq p \leq n/2 + 1$.) If $p \geq 5$, then observe that the path $(v_4, v_5, \ldots, v_p, v_1, v_2, v_{p+1}, v_{p+2}, \ldots, v_n)$ cannot be extended to a Hamiltonian path since $v_3 v_n \notin E(G)$, which is a contradiction. Therefore, $p \in \{2, 3, 4\}$. Of course, $G = C_n$ if $p = 2$. If $p = 3$, then we see that $v_1 v_4 \in E(G)$ by considering the path $(v_2, v_3, v_5, v_6, \ldots, v_n, v_1)$. Thus, G contains every possible edge by Lemmas 3.3 and 3.4, that is, G is complete.

Suppose finally that $p = 4$. If n is odd, say $n = 2k + 1$ for some integer $k \geq 3$, then one of $v_1 v_{k+1}$ and $v_1 v_{k+2}$ belongs to $E(G)$ by Lemma 3.4. Then Lemma 3.3 guarantees that $v_1 v_{k+1}, v_1 v_{k+2} \in E_k \subseteq E(G)$, which implies that $v_1 v_{n-2} \in E(G)$ by Lemma 3.4. However, this is impossible since $p \neq 3$. Thus, n must be even and G contains every possible edge without making odd cycles. That is, G is a regular complete bipartite graph. $\square$

A graph G is *strongly randomly traceable* if for every two distinct vertices u and v in G, a Hamiltonian $u - v$ path always results upon starting at u and successively proceeding to any vertex not yet encountered with the added restriction that v is not to be taken should some other eligible vertex still be available.

Theorem 3.55 ([16]). *A graph G of order at least 3 is strongly randomly traceable if and only if G is complete.*

Fig. 3.6 The Coxeter graph

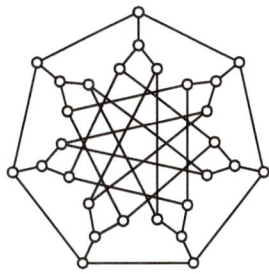

Traceably Singular Graphs That Are Not Traceable

The examples of traceably singular graphs we have seen thus far are those that are homogeneously traceable. Are there graphs that are traceably singular and not traceable? In other words, are there graphs G for which $t(G) = t^\star(G) \geq |V(G)|$? Note that if a graph is not traceable, then it is neither Hamiltonian nor hypohamiltonian.

A graph G is *vertex-transitive* if for every pair u, v of vertices of G, there exists an automorphism $\phi : V(G) \to V(G)$ such that $\phi(u) = v$. For example, the Petersen graph, complete graphs, and regular complete bipartite graphs are vertex-transitive. Certainly, every vertex-transitive graph is regular, but not conversely. Suppose that G is a nontrivial connected vertex-transitive graph and let $v \in V(G)$ with $t(v) = t(G)$. Then G contains a spanning walk $(v = v_0, v_1, \ldots, v_{t(G)})$ of length $t(G)$ whose initial vertex is v. Let u be a vertex in G distinct from v and $\phi : V(G) \to V(G)$ an automorphism such that $u = \phi(v) = \phi(v_0)$. Then G contains a spanning walk $(u = \phi(v_0), \phi(v_1), \ldots, \phi(v_{t(G)}))$ of length $t(G)$ whose initial vertex is u. That is, $t(v) = t(G)$ for all $v \in V(G)$. As a consequence, we obtain the following.

Theorem 3.56. *Every connected vertex-transitive graph is traceably singular.*

Of course, the converse of Theorem 3.56 does not hold. For example, every Hamiltonian graph that is not regular is traceably singular but not vertex-transitive.

There are currently only five known nontrivial connected vertex-transitive graphs that are not Hamiltonian, namely K_2, the Petersen graph (order 10), the *Coxeter graph* (order 28, see Fig. 3.6), and two graphs derived from the Petersen graph and Coxeter graphs by replacing each vertex by a triangle (see Fig. 3.7). The Petersen graph and Coxeter graph are hypohamiltonian while the last two are called the *truncated Petersen graph* (order 30) and *truncated Coxeter graph* (order 84), respectively, and neither is hypohamiltonian. Each of these five vertex-transitive graphs fails to contain a Hamiltonian cycle; yet all five contain Hamiltonian paths. Therefore, these are examples of non-Hamiltonian homogeneously traceable graphs. It is not known whether there exists a vertex-transitive graph that is not traceable. That is, we know of no vertex-transitive graph G for which

Fig. 3.7 Replacing a vertex
by a triangle

Fig. 3.8 The Zamfirescu
graph of order 36

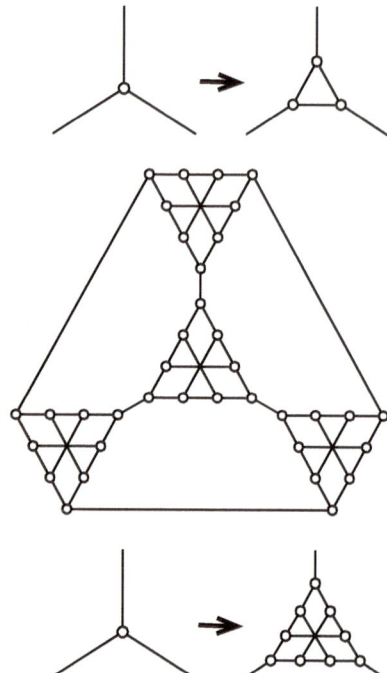

Fig. 3.9 Replacing a vertex
by $P - v$

$t(G) = t^\star(G) \geq |V(G)|$. In fact, Royle conjectured that all other connected vertex-transitive graphs are Hamiltonian (see [24]).

There does exist a traceably singular graph that is neither traceable nor vertex-transitive. The Zamfirescu graph G of order 36, found by Zamfirescu [71], shown in Fig. 3.8, is not vertex-transitive and $t(G) = t^\star(G) = 37$, which was observed by Allen Schwenk (personal communication). The Zamfirescu graph of order 36 is a *snark* (a connected bridgeless cubic graph having chromatic index 4 and girth at least 5, where the *girth* of a graph is the length of a shortest cycle in that graph). Every snark is a non-Hamiltonian graph containing the Petersen graph as a minor while many known snarks are hypohamiltonian, which implies that many known snarks are homogeneously traceable. The Zamfirescu graph of order 36 therefore has a somewhat surprising property of being a non-traceable snark that is still traceably singular.

The Zamfirescu graph of order 36 is obtained by replacing each vertex of a complete graph of order 4 by the graph $P - v$ of order 9 (the Petersen graph P with a vertex deleted), as shown in Fig. 3.9. In general, this operation transforms a cubic Hamiltonian graph of order $2n$ (≥ 4) into another cubic graph G of order $18n$. One may verify that the resulting G is a traceably singular graph with $t(G) = t^\star(G) = 20n - 3$. Clearly G is non-traceable; the length of a longest path in this graph is $\text{diam}_D(G) = 16n + 1$.

3.7   The Total Traceable Number of a Graph

For a nontrivial connected graph G of order n, the *total traceable number* $tt(G)$ is defined by

$$tt(G) = \sum_{v \in V(G)} t(v).$$

Since $t(v) \geq n - 1$ for every $v \in V(G)$, it follows that $tt(G) \geq n(n - 1)$. Furthermore, $tt(G) = n(n - 1)$ if and only if G is homogeneously traceable. Therefore, the total traceable number of a connected graph G of order n can be considered as a measure of how close G is to being homogeneously traceable – the closer $tt(G)$ is to $n(n - 1)$, the closer G is to being homogeneously traceable.

It is immediate that $nt(G) \leq tt(G) \leq nt^{\star}(G)$. By Theorem 3.44, recall that

$$\mathscr{T}(G) = \{t(v) : v \in V(G)\} = \{t(G), t(G) + 1, \ldots, t^{\star}(G)\}.$$

Furthermore, if s is a linear ordering of $V(G)$ for which $d(s) = t(G)$, then $t(v) = t(G)$ when v is either the initial term or the terminal term of s.

Theorem 3.57. *If G is a nontrivial connected graph of order n, then*

$$nt(G) + \binom{t^{\star}(G) - t(G) + 1}{2} \leq tt(G) \leq nt^{\star}(G) - \left(\binom{t^{\star}(G) - t(G) + 1}{2} + t^{\star}(G) - t(G) \right).$$

Proof. Since the result is obvious when $t(G) = t^{\star}(G)$, we may assume that $t(G) < t^{\star}(G)$. Let us first consider the lower bound. By Theorem 3.44, there is at least one vertex v for which $t(v) = t(G) + i$ for $1 \leq i \leq t^{\star}(G) - t(G)$ and at most $n - t^{\star}(G) + t(G)$ vertices for which $t(v) = t(G)$. Thus, $tt(G) \geq nt(G) + (1 + 2 + \cdots + (t^{\star}(G) - t(G)))$, as desired. The upper bound can be verified in a similar manner, remembering that there are at least two vertices having traceable number $t(G)$. $\qquad \square$

As for other Hamiltonian numbers and traceable numbers, the total traceable number of a connected graph is bounded above by the total traceable number of a connected spanning subgraph of that graph.

Theorem 3.58. *If H is a connected spanning subgraph of a nontrivial graph G, then $tt(G) \leq tt(H)$. In particular,*

$$tt(G) \leq \min\{tt(T) : T \text{ is a spanning tree of } G\}.$$

Hence, it is useful to know the total traceable numbers of trees. In order to do this, we first recall a useful result from Sect. 3.5 on the traceable number of a vertex of a tree; that is, if T is a nontrivial tree of order n, then

$$t(v) = 2(n - 1) - e(v) \tag{3.4}$$

for every $v \in V(T)$. Using (3.4), it is straightforward to verify that

$$tt(T) = n(n - 1) + \begin{cases} \lfloor (n - 1)^2/4 \rfloor & \text{if } T \text{ is a path of order } n \geq 2 \\ n^2 - 3n + 1 & \text{if } T \text{ is a star of order } n \geq 3 \\ n^2 - 4n + 2 & \text{if } T \text{ is a double star of order } n \geq 4. \end{cases}$$

If $\mathscr{S}_n$ denotes the set of trees of order n, then

$$\{tt(T) : T \in \mathscr{S}_2\} = \{2\}$$

$$\{tt(T) : T \in \mathscr{S}_3\} = \{7\}$$

$$\{tt(T) : T \in \mathscr{S}_4\} = \{14\} \cup \{17\}$$

$$\{tt(T) : T \in \mathscr{S}_5\} = \{24\} \cup \{27\} \cup \{31\}$$

$$\{tt(T) : T \in \mathscr{S}_6\} = \{36\} \cup \{40, 41\} \cup \{44\} \cup \{49\}$$

$$\{tt(T) : T \in \mathscr{S}_7\} = \{51\} \cup \{55, 56\} \cup \{60, 61, 62\} \cup \{65\} \cup \{71\}.$$

For a nontrivial tree T, there is exactly one central vertex if $\operatorname{diam}(T)$ is even and there are exactly two central vertices otherwise. This gives us sharp upper and lower bounds for the total traceable numbers of trees in terms of order and diameter.

Theorem 3.59. *Let T be a tree of order $n \geq 3$ and diameter d. Then*

$$n(n - 1) + b' \leq tt(T) \leq n(n - 1) + b'',$$

where b' and b'' are the integers given by

$$b' = n^2 - (d + 1)n + \lfloor d/2 \rfloor \lceil d/2 \rceil$$

$$b'' = n^2 - \lfloor (d + 5)/2 \rfloor - \lfloor d/2 \rfloor^2 + \lceil d/2 \rceil + 1.$$

Proof. We consider two cases according to the parity of d.

Case 1. d is even. Then $d = 2r$, where $r = \operatorname{rad}(T)$. In this case, observe that $b' = n^2 - (2r + 1)n + r^2$ and $b'' = n^2 - (r + 2)n - r^2 + r + 1$. Thus, T contains exactly one vertex with eccentricity r and at least two vertices with eccentricity i for each i with $r + 1 \leq i \leq 2r$. Therefore,

$$\sum_{v \in V(T)} e(v) \leq r + 2 \left(r^2 + \binom{r+1}{2} \right) + (n - 2r - 1)(2r) = 2rn - r^2$$

while

$$\sum_{v \in V(T)} e(v) \geq r + 2\left(r^2 + \binom{r+1}{2}\right) + (n - 2r - 1)(r + 1) = (r + 1)n + r^2 - r - 1.$$

Since $tt(T) = n(2(n - 1)) - \sum_{v \in V(T)} e(v)$ by (3.4), it follows that

$$n(n - 1) + n^2 - (2r + 1)n - r^2 \leq tt(T) \leq n(n - 1) + n^2 - (r + 2)n - r^2 + r + 1,$$

that is, $n(n - 1) + b' \leq tt(T) \leq n(n - 1) + b''$.

Case 2. d is odd. Then $d = 2r - 1$, where $r = \mathrm{rad}(T)$. In this case, we have $b' = n^2 - 2rn + r(r - 1)$ and $b'' = n^2 - (r + 2)n - r(r - 3)$. Hence, T contains exactly two vertices with eccentricity r and at least two vertices with eccentricity i for each i with $r + 1 \leq i \leq 2r - 1$. Then one can verify that

$$(r + 1)n + r(r - 3) \leq \sum_{v \in V(T)} e(v) \leq (2r - 1)n - r(r - 1).$$

Therefore, the desired result is again obtained.   □

For a fixed integer $d \geq 2$, let $P = (v_0, v_1, \ldots, v_d)$ be a path of length d. If T is a tree of order n and diameter d, then $tt(T) = n(n - 1) + b'$ if and only if T is obtained from P by adding $n - d - 1$ new vertices and joining each of these vertices to either v_1 or v_{d-1}. On the other hand, $tt(T) = n(n - 1) + b''$ if and only if T is obtained from P by adding $n - d - 1$ new vertices and joining each of these vertices to exactly one of $v_{\lfloor d/2 \rfloor}$ and $v_{\lceil d/2 \rceil}$.

In general, if b is an integer such that $b' \leq b \leq b''$, then a tree of order n and diameter d can be constructed from the path P by adding pendant edges at appropriate vertices of P such that its total traceable number equals $n(n - 1) + b$. Thus, we have the following realization result.

Theorem 3.60. *Let n, d, and a be integers such that $2 \leq d \leq n - 1$. Then there exists a tree T of order n, $\mathrm{diam}(T) = d$, and $tt(T) = a$ if and only if $a = n(n - 1) + b$ for some integer b satisfying*

$$n^2 - (d + 1)n + \lfloor d/2 \rfloor \lceil d/2 \rceil \leq b \leq n^2 - \lfloor (d + 5)/2 \rfloor n - \lfloor d/2 \rfloor^2 + \lceil d/2 \rceil + 1.$$

Consequently, if T is neither a star nor a double star, then

$$n(n - 1) + \lfloor (n - 1)^2/4 \rfloor \leq tt(T) \leq n(n - 1) + (n^2 - 4n - 1).$$

Let G be a connected graph of order $n \geq 3$. Recall that the detour diameter $\mathrm{diam}_D(G)$ is the length of a longest path in G and so $2 \leq \mathrm{diam}_D(G) \leq n-1$. While $\mathrm{diam}(T) = \mathrm{diam}_D(T)$ when T is a tree, $\mathrm{diam}(G) \leq \mathrm{diam}_D(G)$ for a connected graph G in general. If G is connected, then it is not always the case that G contains a spanning tree T such that $\text{diam}(T) = \text{diam}(G)$. However, the following always holds.

Lemma 3.5. *If P is a longest path in a connected graph G, then G has a spanning tree containing P. In other words, every connected graph G contains a spanning tree T such that* $\text{diam}(T) = \text{diam}_D(G)$.

Proof. We proceed by induction on the order of G. If G is a connected graph of order at most 4, then either $G = K_{1,3}$ or G is traceable. Let $n \geq 4$ be an integer and suppose that every connected graph of order n has the desired property. Now assume that G is a connected graph of order $n + 1$ and let P be a longest path in G. Since the desired result is immediate if $V(G) = V(P)$, let us assume that the set $S = V(G) - V(P)$ is nonempty. Select $x \in S$ such that $d(x, V(P)) = \max\{d(v, V(P)) : v \in S\}$. Hence, x is not a cut-vertex. Let $y \in V(P)$ such that $d(x, y) = d(x, V(P))$ and consider an $x - y$ geodesic $Q = (x = v_0, v_1, \ldots, v_\ell = y)$. (Since $x \notin V(P)$, observe that $\ell = d(x, y) \geq 1$.) Now consider the graph $H = G - x$. Observe that H is a connected graph of order n containing P, so there exists a spanning tree T_H of H containing P. Then by adding a pendant edge at v_1 in T_H, we obtain a spanning tree of G containing P as well.   □

We have seen that $\text{diam}_D(G) = 2$ if and only if G is a star and $\text{diam}_D(G) = 3$ if and only if G is either a double star or $G = K_{1,n-1} + e$. It can be easily verified that $tt(K_{1,n-1} + e) = n(n-1) + (n^2 - 4n + 1)$. Therefore, if G is a connected graph of order $n \geq 5$ and $\text{diam}_D(G) \geq 4$, then $n(n-1) \leq tt(G) \leq n(n-1) + (n^2 - 4n - 1)$ by Theorem 3.60 and Lemma 3.5.

Let us now turn our attention to those pairs n, a of integers such that $n \geq 5$ and

$$n(n-1) + \lfloor (n-1)^2/4 \rfloor \leq a \leq n(n-1) + (n^2 - 4n - 1) \qquad (3.5)$$

for which there exists a connected graph G of order n and $tt(G) = a$.

For integers n and d with $n \geq 6$ and $5 \leq d \leq n - 1$, let $b'_{n,d-1}$ and $b''_{n,d}$ be the integers given by

$$b'_{n,d-1} = n^2 - dn + \lfloor (d-1)/2 \rfloor \lceil (d-1)/2 \rceil$$
$$b''_{n,d} = n^2 - \lfloor (d+5)/2 \rfloor n - \lfloor d/2 \rfloor^2 + \lceil d/2 \rceil + 1.$$

Define $g_n : \{5, 6, \ldots, n - 1\} \to \mathbb{Z}$ by

$$g_n(d) = \max\{0, b'_{n,d-1} - b''_{n,d} - 1\}. \qquad (3.6)$$

That is, the value of $g_n(d)$ gives us the number of integers "missing" between the total traceable numbers of trees of diameters d and $d - 1$. For example,

Fig. 3.10 Graphs G_1, G_2,
and G_3

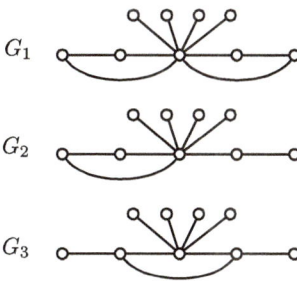

$$\{tt(T) : T \in \mathscr{S}_8, \text{diam}(T) \geq 4\} = \{68\} \cup \{73, 74, 75\} \cup \{78, 79, 80\}$$

$$\cup \{84, 85, 86, 87\}.$$

Since

$$\{tt(T) : T \in \mathscr{S}_8, \text{diam}(T) = 7\} = \{68\}$$

$$\{tt(T) : T \in \mathscr{S}_8, \text{diam}(T) = 6\} = \{73, 74, 75\},$$

there are $g_8(7) = 4$ "missing" integers (namely, 69, 70, 71, and 72) between the two sets.

Theorem 3.61. *For integers n, d with $5 \leq d \leq n - 1$,*
(a) $g_n(5) = 3$,
(b) $g_n(n - 1) = \lfloor n/2 \rfloor$,
(c) $g_n(n - 2) = 2$ *if $n \geq 8$ and is even, $g_n(n - 2) = 3$ if $n \geq 7$ and is odd, and*
(d) $g_9(6) = g_{11}(7) = g_{12}(9) = 1$ *and $g_{10}(7) = 2$.*
Otherwise, $g_n(d) = 0$.

In [55], how those "gaps" due to the values of $g_n(d)$ being nonzero can be filled in is described. The graphs G_i ($1 \leq i \leq 3$) in Fig. 3.10 are obtained from a path P of order 5 by adding $n - 5$ pendant edges at the central vertex of P and adding a few additional edges as shown. Observe then that $tt(G_i) = n(n - 1) + (n^2 - 5n) + i = n(n - 1) + b''_{n,5} + i$. These three graphs correspond to $g_n(5) = 3$.
For $g_n(n - 1) = \lfloor n/2 \rfloor$, assume first that n is even. Let $P = (v_1, v_2, \ldots, v_{n-2})$ and $Q = (w_1, w_2)$ be vertex-disjoint paths of orders $n - 2$ and 2, respectively. Let G_1 be the graph obtained from P and Q by adding the edges $w_i v_2$ ($i = 1, 2$) and $v_{n/2-2}v_{n/2}$. For each integer j with $1 \leq j \leq \lfloor n/4 \rfloor$, construct G_{2j} from P and Q by adding the edges $w_i v_{j+1}$ ($i = 1, 2$). Also, for each integer j with $1 \leq j \leq \lfloor (n - 2)/4 \rfloor$, construct G_{2j+1} from P and Q by adding the edges $w_i v_{j+1}$ ($i = 1, 2$) and $w_2 v_{j+2}$ and deleting the edge $w_1 w_2$. Then we have constructed the graphs $G_1, G_2, \ldots, G_{n/2}$ and

Fig. 3.11 Graphs G_1, $G_2, \ldots, G_5$ for $n = 10$

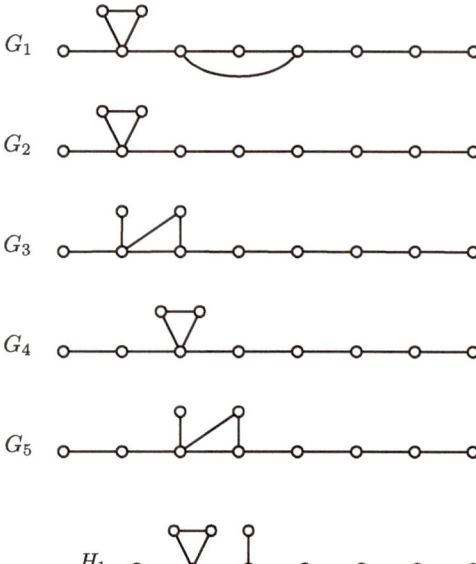

Fig. 3.12 Graphs H_1 and H_2 for $n = 10$

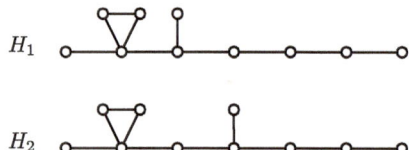

$$tt(G_i) = n(n-1) + \lfloor (n-1)^2/4 \rfloor + i = n(n-1) + b''_{n,n-1} + i$$

for $1 \leq i \leq \lfloor n/2 \rfloor$. A similar construction exists when n is odd. Figure 3.11 shows the graphs $G_1, G_2, \ldots, G_5$ for $n = 10$.

Next assume that $n \geq 8$ and n is even. Thus, $g_n(n-2) = 2$. Let $P = (v_1, v_2, \ldots, v_{n-3})$ and $Q = (w_1, w_2)$ be vertex-disjoint paths of orders $n - 3$ and 2, respectively. For $i = 1, 2$, construct H_i from P and Q by adding the edges $w_j v_2$ $(j = 1, 2)$ and a pendant edge at the vertex $v_{n/2-4+i}$. Then $tt(H_i) = n(n-1) + b''_{n,n-2} + i$ for $i = 1, 2$. See Fig. 3.12 for $n = 10$. When n is odd, a similar construction produces three graphs corresponding to $g_n(n-2) = 3$.

Let $G_{n,a}$ be a connected graph of order n with $tt(G_{n,a}) = a$. According to Theorem 3.61(d), we still need to determine the existence of graphs $G_{9,104}$, $G_{10,127}$, $G_{10,128}$, $G_{11,162}$, and $G_{12,183}$. It turns out that all graphs listed above exist and are shown in Fig. 3.13.

Theorem 3.62. *A pair n, a of integers with $n \geq 5$ and*

$$n(n-1) + \lfloor (n-1)^2/4 \rfloor \leq a \leq n(n-1) + (n^2 - 4n - 1)$$

is realizable as the order and total traceable number of some connected graph.

Fig. 3.13 Graphs $G_{n,a}$

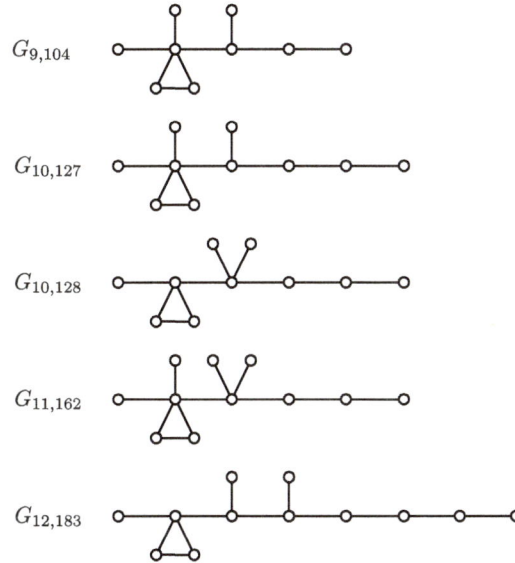

It remains to study those pairs n, a for which $n \geq 5$ and

$$n(n-1) \leq a \leq n(n-1) + \lfloor (n-1)^2/4 \rfloor - 1. \tag{3.7}$$

It is shown in [55] that for every pair n, a satisfying (3.7) there exists a graph whose order and traceable number are n and a, respectively. Although we do not include details here, the idea is as follows. If we let $k_n = \lfloor (n+5)/4 \rfloor$, then it follows that $(k_n - 2)n < \lfloor (n-1)^2/4 \rfloor \leq (k_n - 1)n$. Let k be an integer such that $2 \leq k \leq k_n$.

If ℓ is another integer such that $3 \leq \ell \leq n - k + 1$, then let C be a cycle of order ℓ and $K = K_{n-k-\ell+2} + (k-2)K_1$ such that $V(C) \cap V(K) = \emptyset$. A graph G is constructed by joining every vertex of K to a vertex belonging to C. Then one can verify that $tt(G) = n(n-1) + (k-2)n + \ell - 2$.

Also, for each integer $i = 1, 2, 3$, let $K = K_{n-k+1-i}$ and $P = P_i$ be a complete graph and a path of orders $n - k + 1 - i$ and i, respectively, such that $V(K) \cap V(P) = \emptyset$. Let u and v be two distinct vertices in K. Then obtain the graph G by (i) joining an end-vertex of P to u and (ii) adding $k - 1$ pendant edges at v and observe that $tt(G) = n(n-1) + (k-1)n - k - 1 + i$.

Finally, let $P = (v_1, v_2, \ldots, v_6)$ be a path of order 6 and suppose that $k \geq 3$. For each integer $i \in \{1, 2, \ldots, k-2\}$, let $K = K_i$ and $K' = K_{n-k-4-i}$ be complete graphs of orders i and $n - k - 4 - i$, respectively, such that P, K, and K' are pairwise vertex-disjoint. Then let the graph G of order n obtained from P, K, and K' by (i) joining every vertex of K to both u_3 and u_4, (ii) joining every vertex of K' to both u_2 and u_3, and (iii) adding $k - 2$ pendant edges at u_2 has $tt(H_{k,i}) = n(n-1) + (k-1)n - k + 2 + i$.

Theorem 3.63 ([55]). *Every pair n, a of integers with $n \geq 5$ and*

$$n(n-1) \leq a \leq n(n-1) + \lfloor (n-1)^2/4 \rfloor - 1$$

is realizable as the order and total traceable number of some connected graph.

Combining Theorems 3.62 and 3.63, we obtain a complete realization result.

Theorem 3.64. *A pair n, α of integers with $n \geq 3$ is realizable as the order and total traceable number of some connected graph if and only if*

$$n(n-1) \leq \alpha \leq n(n-1) + (n^2 - 3n + 1)$$

and exactly one of the following (a)–(c) occurs.
(a) $n = 3$.
(b) $n = 4$ *and* $\alpha \in \{12, 13, 14, 17\}$.
(c) $n \geq 5$ *and either*
    i. $\alpha \leq n(n-1) + (n^2 - 4n + 2)$ *and* $\alpha \neq n(n-1) + (n^2 - 4n)$ *or*
    ii. $\alpha = n(n-1) + (n^2 - 3n + 1)$.

References

1. Alspach, B.: Research problems, Problem 3. Discrete Math. **36**, 333 (1981)
2. Alspach, B.: The wonderful Walecki construction. Bull. Inst. Comb. Appl. **52**, 7–20 (2008)
3. Alspach, B., Gavlas, H.: Cycle decompositions of K_n and $K_n - I$. J. Comb. Theory B **81**, 77–99 (2001)
4. Andrews, E., Chartrand, G., Lumduanhom, C., Zhang, P.: On Eulerian walks in graphs. Bull. Inst. Comb. Appl. **68**, 12–26 (2013)
5. Applegate, D., Bixby, R., Chvátal, V., Cook, W.: On the solution of traveling salesman problems. In: Proceedings of the International Congress of Mathematicians. Volume III of Documenta Mathematica, Berlin, pp. 645–656 (1998)
6. Applegate, D., Bixby, R., Chvátal, V., Cook, W.: The Traveling Salesman Problem: A Computational Study. Princeton Series in Applied Mathematics. Princeton University Press, Princeton (2006)
7. Asano, T., Nishizeki, T., Watanabe, T.: An upper bound on the length of a Hamiltonian walk of a maximal planar graph. J. Graph Theory **4**, 315–336 (1980)
8. Asano, T., Nishizeki, T., Watanabe, T.: An approximation algorithm for the Hamiltonian walk problems on maximal planar graphs. Discrete Appl. Math. **5**, 211–222 (1983)
9. Bauer, D., Broersma, H.J., Veldman, H.J.: Not every 2-tough graph is Hamiltonian. Discrete Appl. Math. **99**, 317–321 (2000)
10. Bermond, J.C.: On Hamiltonian walks. Congr. Numer. **15**, 41–51 (1976)
11. Biggs, N.L., Llod, E.K., Wilson, R.J.: Graph Theory 1736–1936. Oxford University Press, New York (2006)
12. Bondy, J.A., Chvátal, V.: A method in graph theory. Discrete Math. **15**, 111–136 (1976)
13. Bondy, J.A., Halverstam, F.Y.: Parity theorems for paths and cycles in graphs. J. Graph Theory **10**, 107–115 (1986)
14. Bryant, D., Horsley, D., Pettersson, W.: Cycle decompositions of complete graphs. Proc. London Math. Soc. (to appear) arXiv:(1204)3709V1 [math.CO] 17 Apr 2012
15. Chartrand, G.: The existence of complete cycles in repeated line-graphs. Bull. Am. Math. Soc. **71**, 668–670 (1965)
16. Chartrand, G., Kronk, H.V.: Randomly traceable graphs. SIAM J. Appl. Math. **16**(4), 696–700 (1968)
17. Chartrand, G., Wall, C.E.: On the Hamiltonian index of a graph. Stud. Sci. Math. Hungar. **8**, 43–48 (1973)
18. Chartrand, G., White, A.T.: Randomly traversable graphs. Elem. Math. **25**, 101–107 (1970)
19. Chartrand, G., Zhang, P.: Chromatic Graph Theory. Chapman & Hall/CRC, Boca Raton (2009)
20. Chartrand, G., Zhang, P.: A First Course in Graph Theory. Dover, New York (2012)
21. Chartrand, G., Polimeni, A.D., Stewart, M.J.: The existence of 1-factors in line graphs, squares, and total graphs. Nedrl. Akad. Wetensch. Proc. Ser. A 76 Indag. Math. **35**, 228–232 (1973)
22. Chartrand, G., Gould, R.J., Kapoor, S.F.: On homogeneously traceable nonhamiltonian graphs. Ann. N. Y. Acad. Sci. **31**, 130–135 (1979)

23. Chartrand, G., Thomas, T., Saenpholphat, V., Zhang, P.: A new look at Hamiltonian walks. Bull. Inst. Comb. Appl. **42**, 37–52 (2004)
24. Chartrand, G., Lesniak, L., Zhang, P.: Graphs & Digraphs, 5th edn. CRC, Boca Raton (2010)
25. Chartrand, G., Jordon, H., Zhang, P.: Edge colorings and decompositions of Eulerian graphs. Australas. J. Comb. **58**, 48–59 (2014)
26. Chvátal, V.: Tough graphs and Hamiltonian circuits. Discrete Math. **5**, 215–228 (1973)
27. Chvátal, V., Erdős, P.: A note on Hamiltonian circuits. Discrete Math. **2**, 111–113 (1972)
28. Cook, W.: In Pursuit of the Traveling Salesman. Mathematics at the Limits of Computation. Princeton University Press, Princeton (2012)
29. Dirac, G.: Some theorems on abstract graphs. Proc. Lond. Math. Soc. **2**, 69–81 (1952)
30. Dirac, G.: On arbitrarily traceable graphs. Math. Scand. **31**, 319–378 (1972)
31. Enomoto, H., Jackson, B., Katerinis, P., Saito, A.: Toughness and the existence of k-factors. J. Graph Theory **9**, 87–95 (1985)
32. Euler, L.: Solutio problematis ad geometriam situs pertinentis. Comment. Academiae Sci. I. Petropolitanae **8**, 128–140 (1736)
33. Fleischner, H.: The square of every two-connected graph is Hamiltonian. J. Comb. Theory B **16**, 29–34 (1974)
34. Fleischner, H.: Eulerian Graphs and Related Topics, part 1, vol. 2. Number 50 in Annals of Discrete Mathematics. North-Holland, Amsterdam (1991)
35. Fujie, F., Zhang, P.: On total distances of cyclic orderings and Hamiltonian spectrums. Congr. Numer. (to appear)
36. Fujie-Okamoto, F.: On traceable and upper traceable numbers of graphs. Ars Comb. (to appear)
37. Goodman, S.E., Hedetniemi, S.T.: On Hamiltonian walks in graphs. SIAM J. Comput. **3**, 214–221 (1974)
38. Guy, R.K.: A Problem of Zarankiewicz. In: Theory of Graphs, 1968 (Proceedings of the Colloquium, Tihany), pp. 119–150. Academic, New York (1966)
39. Harary, F., Nash-Williams, C.S.J.A.: On Eulerian and Hamiltonian graphs and line graphs. Can. Math. Bull. **8**, 701–709 (1965)
40. Heinrich, K., P. Horák, Rosa, A.: On Alspach's conjecture. combinatorial designs – a tribute to Haim Hanani. Discrete Math. **77**, 97–121 (1989)
41. Herz, J.C., Duby, J.J., Vigué, F.: Recherche systématique des graphes hypohamiltoniens. In: Rosenstiehl, P. (ed.) Theory of Graphs: International Symposium, Rome, pp. 153–159. Gordon and Breach, Paris (1966)
42. Hierholzer, C.: Über die möglichkeit, einen Linienzug ohne Wiederholung und ohne Unterbrechnung zu umfahren. Math. Ann. **6**, 30–32 (1873)
43. Jackson, B.: Hamilton cycles in regular 2-connected graphs. J. Comb. Theory B **29**, 27–46 (1980)
44. Kawarabayashi, K., Ozeki, K.: Spanning closed walks and TSP in 3-connected planar graphs. In: Proceedings of the Twenty-Third Annual ACM-SIAM Symposium on Discrete Algorithms (SODA'12), Kyoto, pp. 671–682 (2012)
45. Kirkman, T.P.: On a problem in combinatorics. Camb. Dublin Math. J. **2**, 191–204 (1847)
46. König, D.: Theorie der endlichen und unendlichen Graphen. Akademische Verlagsgesellschaft, Leipzig (1936)
47. Král, D., Tong, L.D., Zhu, X.: Upper Hamiltonian numbers and Hamiltonian spectra of graphs. Australas. J. Comb. **35**, 329–340 (2006)
48. Kwan, M.-K.: Graphic programming using odd or even points. Acta Math. Sinica **10**, 264–266 (Chinese); translated as Chinese Math. **1**, 273–277 (1960)
49. Lindgren, W.F.: An infinite class of hypohamiltonian graphs. Am. Math. Mon. **74**(9), 1087–1089 (1967)
50. Liu, D.: Hamiltonian spectrum for trees. Ars Comb. **99**, 415–419 (2011)
51. McKee, T.A.: Recharacterizing Eulerian: intimations of new duality. Discrete Math. **51**, 237–277 (1984)
52. Nash-Williams, C.S.J.A.: Edge-disjoint Hamiltonian Circuits in Graphs with Vertices of Large Valency. Studies in Pure Mathematics, pp. 157–183. Academic, London (1971)

53. Okamoto, F., Saenpholphat, V., Zhang, P.: Measures of traceability in graphs. Math. Bohem. **131**, 63–83 (2006)
54. Okamoto, F., Saenpholphat, V., Zhang, P.: The upper traceable number of a graph. Czech. Math. J. **58**, 271–287 (2008)
55. Okamoto, F., Zhang, P.: The total traceable number of a graph. Util. Math. **85**, 13–31 (2011)
56. Ore, O.: Note on Hamiltonian circuits. Am. Math. Mon. **67**, 55 (1960)
57. Ore, O.: Thoery of Graphs. American Mathematical Society Colloquium Publications, Providence (1962)
58. Ore, O.: Hamiltonian connected graphs. J. Math. Pures Appl. **42**, 21–27 (1963)
59. Saenpholphat, V., Zhang, P.: Graphs with prescribed order and Hamiltonian number. Congr. Numer. **175**, 161–173 (2005)
60. Šajna, M.: Cycle decompositions III: Complete graphs and fixed length cycles. J. Comb. Des. **10**, 27–78 (2002)
61. Sekanina, M.: On an ordering of the set of vertices of a connected graph. Spisy Přírod Fak. Univ. Brno. **412**, 137–141 (1960)
62. Shank, H.: Some parity results on binary vector spaces. Ars Comb. **8**, 107–108 (1979)
63. Skupień, Z.: Homogeneously traceable and Hamiltonian connected graphs. Demonstratio Math. **17**, 1051–1067 (1984)
64. Toida, S.: Properties of an Euler graph. J. Franklin Inst. **295**, 343–345 (1973)
65. Truszczyński, M.: Centers and centroids of unicyclic graphs. Math. Slovaca **35**, 223–228 (1985)
66. Tutte, W.T.: The dissection of equilateral triangles into equilateral triangles. Proc. Camb. Philos. Soc. **44**, 463–482 (1948)
67. Tutte, W.T., Smith, C.A.B.: On unicursal paths in a network of degree 4. Am. Math. Mon. **48**, 233–237 (1941)
68. van Aardenne-Ehrenfest, T., de Bruijn, N.G.: Circuits and trees in oriented linear graphs. Simon Stevin **28**, 203–217 (1951)
69. Veblen, O.: An application of modular equations in analysis situs. Ann. Math. **14**, 86–94 (1912)
70. Veblen, O.: Analysis Situs. American Mathematical Society Colloquium Lectures, vol. 5. American Mathematical Society, New York (1922)
71. Zamfirescu, T.I.: On longest paths and circuits in graphs. Math. Scand. **38**, 211–239 (1976)
72. Zhu, Y., Liu, Z., Yu, Z.: An improvement of Jackson's result on Hamiltonian cycles in 2-connected regular graphs. Ann. Discrete Math. **115**(27), 237–247 (1985)

Index